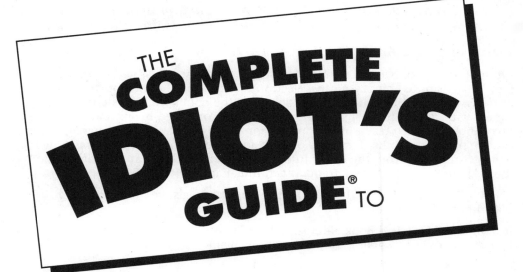

THE **COMPLETE IDIOT'S GUIDE®** TO

Life Science

by Lesley A. DuTemple

alpha books

Macmillan USA, Inc.
201 West 103rd Street
Indianapolis, IN 46290

A Pearson Education Company

For my family, each and every one of you—past, present, and future. May the mighty river of love, and that frail thread of DNA, always connect us.

International Standard Book Number: 0-02-863199-4
Library of Congress Catalog Card Number: Available upon request.

02   01   00   8   7   6   5   4   3   2   1

Interpretation of the printing code: The rightmost number of the first series of numbers is the year of the book's printing; the rightmost number of the second series of numbers is the number of the book's printing. For example, a printing code of 00-1 shows that the first printing occurred in 2000.

Printed in the United States of America

Publisher
Marie Butler-Knight

Product Manager
Phil Kitchel

Managing Editor
Cari Luna

Acquisitions Editor
Randy Ladenheim-Gil

Development Editor
Suzanne LeVert

Production Editor
Billy Fields

Copy Editor
Krista Hansing

Illustrator
Jody P. Schaeffer

Cover Designers
Mike Freeland
Kevin Spear

Book Designers
Scott Cook and Amy Adams of DesignLab

Indexer
Tonya Heard

Layout/Proofreading
Darin Crone
Terri Edwards
Donna Martin
Julie Swenson

Contents at a Glance

Contents

Foreword

The most exciting thing about biology in this day and age is the rapid progress that we are making in understanding our natural world. The results of our new knowledge are already making themselves known in the form of more nutritious foodstuffs, innovative treatments for diseases, and a better understanding of how ecological systems work. Whether or not you develop a professional interest in biology, you may well wish to be clued in as the public debate addresses a huge variety of issues growing out of our new knowledge in biology.

The questions are only just beginning. How should genetic engineering be applied to improve foods and develop new pharmaceuticals? Should the remaining carefully stored stocks of the smallpox virus be obliterated or saved? How will you balance your interest in protecting your medical privacy against the improved medical care that will come of knowing your complete genome? What will stem cells mean to your health as you age?

For those who haven't made a specialty of biology, the field can seem overwhelming with all of its specialized terms. Indeed, the lingo of biology, with many of its terms derived from ancient Greek, Latin, and even modern comic strips, can seem as uncommunicative as a foreign language. But as you gain familiarity with the terms, you'll begin to see, if not method to the madness, at least some entertainment value. Just like the best mnemonic tricks, biological nomenclatures carry hints about the significance of the observation, its most notable feature, who the discoverer was, or where the observation was made.

The idiosyncratic nature of the terminology reflects the unruly pace of discoveries in biology. Often significant observations are made in isolation from other knowledge that could help make sense of the bigger picture. But as research progresses, and each bit of knowledge is used to push for answers to related questions, what at first seems unconnected begins to link up to observations in other arenas. The results can be quite surprising, as we begin to see commonalities in the physiology of animals, plants, and microbes.

And this is where biology begins to make sense. Although you might expect some similarities between a panda bear and a grizzly bear, you might be surprised to learn of the similarity between the messed-up daily cycle of a jet-lagged traveler and the response of plants to changes in day length. Even the features that distinguish one organism from another begin to make sense as you understand the biological building blocks involved. Thus, to learn more about any biological organism is often to learn more about all organisms.

This book will give you a general overview of the entire field of biology. Use it as your only source, or use it as a portal into some of its interesting stories, but either way, you'll learn things that will deepen your understanding of your own world and that can put into context some of the new information about discoveries and threats to our health and ecosystem that you hear about in the news.

Welcome to the wild and wonderful world of biology.

Pamela J. Hines, Ph.D., is a Senior Editor at *SCIENCE* magazine, a weekly journal published for the professional scientist and interested public by the American Association for the Advancement of Science. With a bachelor's degree from Oberlin College and a doctorate from the Johns Hopkins University, Dr. Hines has done research in molecular and developmental biology. She has long been fascinated by the biological world. She and her husband live in Arlington, VA.

Introduction

What is life? Philosophers ponder this question all the time, as do college students who want to avoid studying. But it's not a trick or existentialist question. It's a straightforward question about life: *What is it?*

No one has yet answered the existentialist part, but over the centuries many people have come up with some great (and not so great) answers as to what life is. In fact, there are so many answers and questions that an entire field of science sprang up around the seeking of those answers: life science.

Life science is the study of all living things. But *study* can be such a stern word, and life science is anything but stern. It's the most fascinating field you'll ever encounter. Forget the Starship Enterprise. You want bizarre life forms? Take a look in your own backyard. This planet is full of the most amazing life forms you could ever imagine.

To really understand life science, or biology, you need only two things: your curiosity and your sense of wonder. Those two things will get you farther along the path of life science than anything else ever will. I guarantee you that, with only curiosity and wonder, you'll start to really understand how life science works.

And the best part is, it's all right here. The most amazing things ever seen in this entire universe are *right here*. Think about it: Did anyone ever come back from space oohing and aahing about all the cool stuff they had seen? They ooh and ahh about how cool the *Earth* looks from outer space, but they're remarkably silent about all the cool organisms on the moon (because there simply aren't any).

There are millions and millions of organisms on Earth. No life science book could cover them all. (And as brilliant as I and the editors are, this book can't do it, either.) What this book *can* do is give you a look at the big picture and fire your own curiosity. Here's a brief overview of some of the topics we'll be exploring:

Part 1, "Life: It's All Around You," will provide you with the basics about what it means to be alive. You may intuitively know what's alive and what's not, but as far as science is concerned, there are some real criteria for determining whether an organism is alive. Part 1 also gives you an overview of the development of the general field of science, as well as a brief introduction into matter and the elements.

Part 2, "The Kingdoms of Life," covers the five general classifications for all living organisms. Think of it as a place for everything, and everything in its place. The Kingdom Monera includes bacteria and viruses; the Kingdom Protista includes protists (wonderful squishy creatures) and algae; the Kingdom Fungi is made up of fungi; the Kingdom Plantae includes all plants; and the Kingdom Animalia includes all animals.

Part 3, "How Animals (Human and Otherwise) Live," gives you an overview of the main systems in animals and tells how they interconnect and function together to keep an organism alive. Eating, breathing, exercising, and fighting off the common cold—it's all here (even sex).

Part 4, "How Plants Live," provides you with an overview of the plant kingdom. We'll look at the different types of plants (there's no such thing as a typical plant) and how they adapt to their environments. We'll also look at the distinctions between herbaceous plants (something like a daisy) and woody plants (something like an oak tree), as well as their common ground.

Part 5, "Nature: Always Moving Forward," attempts to tackle the big topics of evolution, genetics, modern genetic engineering, and ecology—how it all fits together.

Extras

We've also added sidebars to give you some help with the terminology (you mean, you don't know what a coacervate is?) as well as to provide you with interesting tidbits, quirky facts, and helpful tips. Here's what you'll find:

Think About It!

These sidebars provide you with some food for thought about the world of science and your place within it.

Try It Yourself

Here you'll find experiments for you to perform that will help make sense out of the mystery of science.

Bio Buzz

These sidebars define unfamiliar words and terms so that you'll be better able to follow the text.

Weird Science

In these sidebars, you'll find lots of quirky tidbits and fascinating facts about the wide world of biology and science.

Acknowledgments

No book ever reaches the reader without the help and input of many people. Special thanks to Lettie Lee and Andree Abecassis, my agents at the Ann Elmo Agency; Randy Ladenheim-Gil and Suzanne LeVert, my editors at Macmillan, who demonstrated amazing serenity in the face of any and everything; and to everyone else at Macmillan and Alpha Books that I've never met, but without whom this book wouldn't be in readers' hands right now. Thanks also to Jim, Meg, and Chris—my beloveds who support me in everything.

Special Thanks to the Technical Reviewer

The Complete Idiot's Guide to Life Science was reviewed by an expert who double-checked the accuracy of what you'll learn here, to help us ensure that this book gives you everything you need to know about biology and the world of science. Special thanks are extended to Dr. Annette Halpern Hinds, an expert in many fields. (She can even fit a kangaroo with a mask for measuring aerobic respiration—she made 'em herself.) Special thanks and appreciation is extended to Annette.

Trademarks

All terms mentioned in this book that are known to be or are suspected of being trademarks or service marks have been appropriately capitalized. Alpha Books and Macmillan USA, Inc., cannot attest to the accuracy of this information. Use of a term in this book should not be regarded as affecting the validity of any trademark or service mark.

Part 1
Life: It's All Around You

Welcome to "Life Science," the most pertinent and magical science you'll ever discover. And that's because it concerns the most amazing thing on this planet—life—and that includes you.

Where did we come from? Where are we going? And for that matter (to abandon the narcissistic viewpoint), where did that gopher, that butterfly, and that violet come from? How would you identify something that's alive? If you whistled and it came, would that mean it's alive? Why? Most of the time, you intuitively recognize what's alive and what's not. You're alive, the rock isn't. But some things that we take for granted are the product of hundreds of years of groping toward answers.

Centuries ago, people looked around them—at all the amazing things—and thought it must be magic. Trying to understand how the magic worked is how science began. And that's what we introduce you to in Part I. All you need is your curiosity and we'll do the rest!

Chapter 1

Science   Life

Science and Life

In This Chapter

➤ What does it mean to be alive?

➤ By magic and experiments: how science developed

➤ Understanding the scientific method

➤ Matter and the elements

Biology. Science. The very words cause most people's eyes to glaze over. Scientific formulas, compound equations, the table of elements. It's enough to scare off the best of us.

But biology, or life science, is more than mind-boggling formulas and equations. In fact, from a purely narcissistic standpoint, it's the most interesting thing around. It's all about *you*, and what could be more interesting than that? (Even your mother couldn't argue with that one.) Biology is the science that deals with plants and animals—in fact, it's all about everything about plants and animals, including their origins, history, characteristics, habits, you name it. Biology also involves how it all fits together. Biology is you and everything around you. It's life.

It's also a lot of ground to cover. To make it easier, we'll break it down into manageable topics. In this chapter, we're going to talk about how scientists determine whether something is alive—a determination that seems self-evident to us when we're dealing with recognizable plants and animals, but not so self-evident when you're dealing with unrecognizable objects, particularly at the cellular level.

We'll also discuss how the general field of science developed and rose out of the realm of magic. As science developed into a true field of its own, the scientific method (a method of testing and verifying questions) also arose. Finally, we'll talk a little bit about the most basic components of life (and nonlife) on Earth: matter and the elements. Again, it's a lot to cover, so let's get started!

A Good Place to Start

Biology is probably the best place to start any preliminary study of science. It's got something for everyone: birds and bees; oozy, slimy fungus; gorgeous flowers and towering trees; cute little furry animals and predatory reptiles; molecules and cells; and blood and guts. Almost any area of science will fit into some category of biology.

Do you like mysteries? The life sciences have kept people wondering for centuries. Just where did that first single-cell creature come from? How fast does evolution work? Are humans more closely related to apes or Neanderthals? How does the brain of a whale work? Which came first, the chicken or the egg? On these questions, your research could probably produce as good of a hypothesis as anyone else's, but scientists all over the world still struggle with them.

Biology even has something for the most contentious among us because it's rife with controversy. Religious philosophers and biologists have been battling it out for the last century, and there appears to be no conclusive end in sight. Even without having to contend with religious viewpoints, the world of biology is still filled with controversy. With new methods of *DNA* identification, for instance, the entire structure of the "tree of life" is under fire, and the fallout from those developments alone will trigger arguments that could go on for centuries.

How could you not like biology, then? It seeps into and fills every pore of your being, philosophically and literally. It's the very essence of life itself!

The Characteristics of Life

If biology is the study of life, a good starting point in understanding it is to figure out exactly what life is. Although you intuitively know what's alive and what's not, for scientific purposes, an object must meet six criteria to be considered alive: movement, organization, homeostasis, energy, reproduction, and growth. Don't worry if you don't understand what we mean by these concepts now—we're about to go through them one by one!

Move, Move, Move!

Movement is probably the most obvious characteristic of life. You've only got to see something move on its own—a rabbit running across a field or a bird flying in the sky—to know that it's alive. You'd be stunned if you saw a rock scooting across the road. You might even blurt out, "Oh my gosh, it's alive!" At first glance, plants aren't always obviously alive (you wouldn't expect a plant to run across the road), but closer observation would show that plants do indeed move. Stems shoot upward, flowers open and close, and leaves follow the movement of the sun, albeit, slowly. For biological purposes, movement is movement, no matter how slow or imperceptible, and that's what counts, biologically speaking.

Organization Is Everything

Anything that's alive usually has a very high level of organization. Things that aren't alive are also organized, but they don't possess nearly the same level of complexity as do living things. A rock, for instance, contains the same minerals in the same proportion as another rock of the same kind. But each rock will be differently arranged. There won't be any rhyme or reason to the organization of one rock to the next because the components of rocks aren't organized into identifiable systems; the composition, arrangement, and shape of each rock always varies.

But for something that's alive—say, a rabbit—that's a different story. The structure and organization of any rabbit will be virtually identical to every other rabbit, and the level of organization will be highly complex. Each rabbit will have a heart, a circulatory system, a respiratory system, digestive organs, and so on. And each system and organ will be in virtually the same place in every single rabbit. Plants also have organs, including their roots, stems, leaves, and flowers, but the arrangement of plant organs are highly varied, a fact we'll discuss in more depth later in the book.

Every organ of a living object is composed of tissues, and all tissues are made up of their own small units, called cells. Most cells are microscopic, but even cells have their own highly organized structure.

Homeo-what?

Another characteristic of life is *homeostasis*, or the tendency to maintain constant conditions within the body. A rabbit, for instance, will maintain the same internal body temperature no matter how hot or cold it is outside. With mammals, like a

Think About It!

Having trouble getting a handle on the topic of life science? Cruise into any bookstore and look under "science." There are plenty of great textbooks out there, but you'll also find some really interesting reads. *Pilgrim at Tinker Creek,* by Annie Dillard, is a fascinating venture into the subject.

Weird Science

All living things have highly organized internal structures. Even the lowliest cell is organized into distinct, functioning parts. Usually, the more complex the organism, the more complex the organization will be. Most mammals, for instance, have such complex organizations that modern science still hasn't been able to decipher much of their brain and nervous system activities.

Bio Buzz

Homeostasis means "remaining the same." It is the process whereby cells regulate and control their own environment.

Weird Science

Plants are the only life forms on Earth that can make food from the sun's energy. The rest of us just use the sun's heat to stay warm. When a plant uses the sun's energy to grow and make food, it's called photosynthesis. *Photo* means "light," and *synthesis* means "to make something."

rabbit, the body does this automatically, but all living things strive toward homeostasis. In hot weather, a rabbit spreads out its ears or keeps them erect in an effort to radiate heat away from the body. In winter, the same rabbit pulls its ears in close to its head in an effort to conserve heat.

Although plants can't regulate their temperature, they close their numerous guard cells in an effort to conserve water and maintain constancy in hot dry weather. Guard cells are special cells found only in plants. By opening and closing, they regulate water loss by the plant. All living things work at maintaining constancy, either deliberately or automatically.

Energy—Use It!

All living things absorb and use energy. Nonliving things can absorb energy, but they don't do anything with it. When the sun shines on a rock, the rock absorbs some heat energy, it isn't capable of doing anything with that energy. When the sun shines on a plant, on the other hand, the plant takes the energy and converts it into the chemical energy of carbohydrates, proteins, and fats, which in turn keep the plant alive and help it produce more roots, stems, and leaves. For animals, plants are a direct source of energy. A rabbit will eat the plant and use the energy to stay alive, run, eat more food, grow new fur, and do any number of other things.

Reproduction

Another trait that living beings share is the ability to reproduce. No one is surprised when two rabbits produce a litter of baby rabbits. Likewise, it isn't surprising when plants reproduce through seeds and other methods. But if two rocks in your garden suddenly produced a bunch of little rocks, you'd be shocked. Only living things can reproduce.

Growing, Growing, Grown

Except for a few cells that reproduce by dividing in half, all living things grow sooner or later. In animals and plants, growth starts with the fertilization of a single egg cell—called an *ovum* in mammals and an *ovule* in plants—and development proceeds from there. During the growth process, cells not only increase in number, but they also develop into the different types of cells that are needed to form the organs and tissues of the new individual. The ability to grow is a basic characteristic of life.

Moving Toward "Real" Science

The very thought of having to study science strikes fear in the hearts of many people, sometimes for good cause. Science can be very complex, but it's not beyond the grasp of most people. Contrary to what most people think, science is neither mysterious nor reserved exclusively for geniuses.

Before embarking on a study of the specific branch of science called biology, it's helpful to take a general look at the world of science, which developed because of a single trait found in all humans: curiosity. It might have killed the cat, but it does wonders for advancing scientific knowledge.

Bio Buzz

In animals, the single egg cell from which growth starts is called an **ovum**. In plants, it is called an **ovule**.

What's That?

Let's take a look at curiosity in its most basic and primitive form. Imagine that you're a cave person—no computer, no CNN, not much of anything but the natural world around you. Future generations will talk about getting in touch with nature, but that's not a problem for you—you're in touch with nature every second. Now imagine that you're walking along (keeping an eye out for saber-toothed tigers), and it starts to rain. It begins to rain harder. Then the lightning starts. The heavens open up and a large bolt of fire streaks earthward. With a mighty *crack*, the tree across the clearing bursts into flame. What do you do?

Well, you're probably scared witless. But if you've got even the glimmerings of human intelligence

Weird Science

Curiosity is the trait most responsible for the advancement of science. Curiosity moves everything. Without it, we wouldn't know how the Earth moves in its orbit, how photosynthesis works, or how snakes reproduce. Curiosity drives humans to find out more.

(and we're going to assume that you do), it won't be long before you're going to want to know just what's going on with that tree. You're *curious*. Your curiosity will overcome your fear, and you'll start looking at the flaming tree, first in amazement and then with increasing puzzlement.

How Does It Work?

Pretty soon, you might notice the fire literally consuming the tree, which was simply vanishing into the flames. You might also notice flames spreading to other places that weren't burning originally, including to another tree. By this time, you're really curious about what's going on here—so curious, in fact, that you take a branch from another tree, walk over to the flaming tree, and see if your branch will catch on fire. You start to *experiment*. Once you start experimenting, it becomes hard to stop. There's always another question to be answered.

7

Before long, you've accumulated a lot of information. Some of your observations become general rules. "Wood burns, but rocks don't," becomes an accepted statement because the same thing always happens, regardless of who performs the experiment. It becomes fact.

Without knowing it, you've just lived science. Curiosity and experimentation are the true foundation stones of science. Thousands of years later, they're still the foundations. Studying science and biology is really just satisfying your natural curiosity.

Is It Science, or Magic?

With curiosity and experimentation forming the cornerstones of science, there was (and is) a lot of leeway for development in the field. The natural world is a big place with lots of goings-on to be observed. It took quite a while for science to develop into the field it is today, and the first efforts toward its study weren't even considered science—they fell under the heading of witchcraft, voodoo, or magic.

Think About It!

Science and biology can always hold your interest. To learn more, check out a copy of J. Moore's book, *Science as a Way of Knowing: The Foundations of Modern Biology* (Harvard University Press, 1993).

Try It Yourself

Energy from the sun is an important part of life. You could even say it's the foundation of life science. Humans certainly revel in it, but for plants, sunlight is not an option. It's so important, in fact, that plants appear to be programmed to reach for sunlight.

To see this in action, try this experiment. Put a houseplant with a sturdy stem (not a vine) in a sunny window. Leave it there for at least a week, preferably of a couple weeks. By all means, water it and take care of it, but otherwise don't move it.

After a couple weeks, pick up the plant, set it on a table, and take a good look at it. It should be lopsided; the entire plant should be leaning in one direction. If you put it back in the window, in the opposite direction, the plant will pull itself up and lean in the other direction. That's the power of sunlight; the moon pulls tides, but the sun pulls plants.

But whether considered witchcraft or a legitimate intellectual pursuit, science still developed through curiosity and observation. Over time, a few people were more observant than others. They watched the natural world carefully, noted correlations between certain events, and drew conclusions from their observations. They knew which substances would spit brilliant colors when thrown on flame, which herbs would cure ailments, and any number of other examples of cause and effect. Their less observant and knowledgeable peers may have considered them to be witch doctors, oracles, and seers, but they were really the first scientists.

Weird Science

Having scientific knowledge can be a dangerous thing. Many early scientists were persecuted or put to death in the name of "witchcraft." In reality, they had just observed and learned more about very natural and earthly things than the average person.

Greek Science

Slowly but surely, man's knowledge of and command over the natural world developed through history. Humans learned to master the use of fire. By mastering fire, they could stay in one place all year—they didn't have to pack up and leave just to stay warm. They learned how to grow and harvest crops, and how to domesticate animals. Growing crops and having your own livestock meant that you didn't have to spend all day searching for food—you had to pay attention to it, but with care, it would be right at your fingertips.

These advances made it possible for people to survive and support themselves, and still have some time left over in the day. And they could spend this time on what we now call leisure activities, including the pursuit of knowledge for knowledge's sake. They were so curious about the world around them that chasing down the answers became their leisure hobby, or even their full-time careers.

Nowhere was this more true than in ancient Greece, which boasted several full-time scientists. Ironically, although the ancient Greeks did much to advance the field of science, they also stalled it in some ways.

Some of the early Greeks, like Aristotle, were careful observers of nature and drew remarkably correct conclusions from their observations. More than 2,000 years ago, Aristotle was the first person to make a distinction between living and nonliving things. He was also first to classify living things as either plants or animals. Aristotle even created subdivisions for animals and plants. He subdivided animals into water, land, or air dwellers and grouped plants by stem differences. As fledgling science, Aristotle's system was definitely on the right track. And it was accurate enough that it was used until the eighteenth century.

But when Euclid managed to develop an entire system of geometry by deducing a few "self-evident" statements, or axioms, that became the standard method of Greek science. Euclid lived in Alexandria, Egypt around 300 B.C.E. and is known as a great mathematics teacher. But he also had a philosophical bent: He loved to ponder and

Weird Science

Although reducing science to the realm of glorious, shining ideas, the early Greek scientists still managed to produce a body of work that has merit today. Euclid's system of geometry did indeed work and *is* still in use (as any high school student can tell you). And Aristotle's methods of observing the natural world set the standards for modern scientists.

Weird Science

For his efforts on behalf of the advancement of science, Galileo was rewarded with a trial by the Inquisition in Rome, was ordered to recant his statements, and was forced to spend the last eight years of his life under house arrest.

think about all the "what-ifs," but he didn't necessarily determine whether they were true. Just by thinking about it, Euclid was able to determine how lines, planes, spheres, and other geometrical shapes worked. He developed theorems, or statements about geometry that can be proven mathematically. That he managed to hit the nail on the head, and that his method really did work fine for geometry was great. But geometry and general science aren't the same thing. Euclid wasn't dealing with life, literally. Nevertheless, Euclid's method of just thinking out the answer and pronouncing it true became the standard. Experimenting and verifying an idea became less important than the idea itself. The imperfect world of real life was not allowed to spoil the perfect world of ideas.

The Greek method of science prevailed for nearly 2,000 years. Until the Renaissance, nearly all scientific questions were answered by "Euclid says …" or "Aristotle says …."

Galileo and the Beginnings of Modern Science

Science progressed with scientists following the Greek method of doing business until the late sixteenth century, when Galileo, an Italian mathematician, astronomer, and physicist, came on the scene. Galileo was such an intellectual powerhouse that he retains the title "The Father of Science" even today.

Galileo was fascinated by how things worked. Unlike previous scientists, figuring out how they worked wasn't enough for him; once he'd cracked the code, he wanted to prove that he was correct. Galileo was the king of scientific experimentation. Once he had an idea, he followed up on every aspect of it. And he had a lot of ideas.

Galileo was the first person to use a telescope and study the heavens, thus establishing that the Earth revolves around the sun rather than the other way around. In building a telescope and proving that the Earth revolved around the sun, his experiments led him to develop the principles of gravity and refraction of light—nearly 100 years before Sir Isaac Newton did. Galileo's accomplishments are extensive.

Wierd Science

Curiosity strikes again! A swinging lamp in the Pita Cathedral drove Galileo to study mathematics and to develop a principle still used today—the principle of the pendulum. While sitting in the cathedral, he observed that the swinging lamp always took the same amount of time to complete an oscillation, no matter how large the range of the swing. Curiosity over this phemenon led him to verify this later in life and to suggest that the same principles could be applied to the regulation of clocks. By coordinating the oscillation with the calibration of the hands, Galileo felt that clock minutes could be regulated because the oscillation would always take the same amount of time. It is this principle on which grandfather, or pendulum, clocks still operate. Until he observed the swinging lamp, Galileo had supposedly received no instruction in mathematics.

But Galileo's biggest contribution was not the discovery of any of these marvelous principles we now find self-evident, but the replacement of Aristotle's nonexperimental approach to science with a rational, mathematical approach. With Galileo, establishing truths through experimentation rather than just deducing through theorizing became the way to pursue science. "All knowledge is vain without the confirmation of experiment," is one of his famous comments.

Nowadays, this axiom is not necessarily true because there are many areas of science in which experiments are virtually impossible to perform. But in most fields of science, you must prove the validity of a hypothesis through extensive, controlled experimentation. Galileo's discipline when it came to verifying through experimentation laid the groundwork for the development of the scientific method.

Think About It!

To learn more about the scientific method, check out S. Carey's book, *A Beginner's Guide to the Scientific Method* (Wadsworth, 1994). It's a good overview of the whole process.

The Scientific Method

Today, science still advances only with the advent of curiosity. Almost immediately upon observing an interesting object or new phenomenon, the scientist—or any curious person—finds questions springing to mind. And once a question is raised, what

11

could be more natural than to look for the answer? Galileo was on the right track in insisting upon experimentation. But nowadays, simple experimentation isn't enough. Instead, scientists seek answers by using the scientific method.

Was one simple experiment ever enough to prove a hypothesis? Nope, not to you and I—but then again, we didn't live in 300 B.C.E.!

The scientific method is a way of posing a question and then rigorously testing it (usually in several different ways) to see whether it is correct. Based on the outcome of each test, the original question will possibly be refined, but it certainly will be tested again. If the idea survives the first test, the next test will be harder. Although the scientific method produces verifiable scientific truths, as you might suspect, it can be very time-consuming. A scientist who follows the scientific method leaves little to chance; he tests every variance over, and over, and over again, devising new tests each time. Experimenting again and again is the name of the game.

The Working Hypothesis

When scientists pose a question, they usually have an answer in mind, but usually that answer is considered tentative until they confirm it by experiment. First, the question is raised (usually due to curiosity), and then a tentative answer to the question is formulated. A tentative answer to a scientific question is called a *working hypothesis*.

Bio Buzz

A **working hypothesis** is the tentative answer to a posed scientific question. It is called a "working" hypothesis to emphasize that the answer is unreliable and still under consideration.

It didn't take long for scientists to realize that the working hypothesis method had some pitfalls—namely, someone working with his or her own idea had a tendency to develop affection for that idea, and thus had a tendency to ignore the particular shortcomings of that idea, even as he or she tried hard to be honest in the pursuit of the truth. To overcome this tendency, scientists learned to develop multiple working hypotheses, which, taken as a group, try to cover all possible answers to the problem.

By using a multiple testing approach, scientists can develop affection for several possible answers or outcomes instead of just one. By using a group of working hypotheses to test an idea, a scientist is more inclined to test and evaluate each of them with more honesty.

Nonchallenging and Challenging Tests

Even with multiple working hypotheses, there's still a danger that a scientist will make the tests too easy. Even the best scientists can succumb to the temptation of setting up an easy test for a hypothesis that doesn't adequately challenge the idea. To overcome this tendency, most scientists use a series of challenging tests rather than a single, nonchallenging test to test the validity of their idea.

In addition, scientists tend to design highly exacting tests to specifically eliminate rather than support hypotheses. The harder a test is to pass, the quicker a scientist can get down to the real nitty-gritty of the hypothesis being tested. When one or more of the working hypotheses survive deliberate attempts to eliminate them, scientists begin to have some genuine confidence in the initial idea.

Unlike the days of Aristotle or Euclid, nowadays no scientific statement is accepted at face value. The scientific method is based on questions and working hypotheses—the more difficult, the better. The initial premise gets tested and retested, usually by different groups of people. Eventually, the idea will be accepted as fact, will be rejected as wrong, or will remain in the gray arena of controversy. It's a slow process, but it's the best method of verification.

Heeere's ... Matter and Elements

Just as biology encompasses all aspects of life, it also encompasses many aspects of general science, including chemistry. The subject of chemistry causes more glazed eyes than even biology does. But (and you're probably not going to like this), it's impossible to study biology without a basic understanding of chemistry. All living things are made out of matter, and matter follows the laws of chemistry.

Matter: Does It Matter?

The material that makes up the Earth, the universe, and everything in it—including you—is called matter. Matter exists in three main forms: solid, liquid, and gas. (There is a fourth form of matter, called plasma, but it exists only in the extremely high temperature of stars.)

Most of the Earth itself is solid, including the crust, continents, rocks, and land. The oceans, rivers, and lakes are liquid. The atmosphere is gas, although microscopic particles of liquid and solid matter are also suspended in the Earth's atmosphere.

But if the pressure or temperature within or around any type of matter changes, the form of the matter will also change.

Bio Buzz

Matter is what a thing is made of. An **atom** is the smallest particle of an element or substance. A **molecule** is the smallest part of an element that can exist by itself and still retain the characteristics of the element or substance.

Take H_2O, for instance. In its solid form, H_2O is ice. But if the temperature increases, ice turns into water, the liquid form of H_2O. If the temperature increases even more, water turns into steam, which is the gaseous form of water.

All matter is capable of assuming these three different states: solid, liquid, and gas. Things we think of as always solid, such as rocks or steel, just need higher temperatures than we normally experience to change states. But given higher temperature or pressure, they, too, will change from solid to liquid to gas.

Matter is composed of tiny, tiny particles called *molecules* and *atoms*. Even with the most powerful microscope available, atoms and molecules are almost impossible to see. But scientists know they exist because their behavior remains constant throughout experiments. (The scientific method strikes again! Even if you can't see something, you can still get answers!)

Molecules and atoms are believed to be in constant motion. Scientists think that all motion would stop at absolute zero (–273° C, or –459.4° F), but this temperature has never been achieved, so it remains only a theory.

In the solid state, molecules remain closely attached to each other, leaving them little room for movement. They're so tightly packed that all molecules can do when they're in a solid form is vibrate. Because the molecules don't really move, then, all solids have a fixed shape and form.

When you add the energy that comes from heat, the molecules begin to vibrate faster and harder. Eventually, they vibrate so fast and hard that they break away from each other and enter the liquid state of matter. In the liquid state, the molecules are free to flow over each other. Matter in a liquid state has a fixed volume but will assume the shape and form of whatever contains it.

Finally, if you keep applying heat and pressure, the molecules move faster and faster until they completely break away from each other and enter the gaseous state. As a gas, the molecules are so far apart that they have neither shape nor volume. A gas will fill whatever space is available.

The Elements: You Are What You Are

All matter, living or not, and in solid, liquid, or gas form, is made up of one or more of only 106 *elements* found on Earth. Ninety-two of these elements occur naturally. The other 14 are man-made, and they break down in a fraction of a second. When an element breaks down, it loses atoms, frequently changing its composition as it does so. The 92 natural elements can also break down, but only over millions of years. For most purposes, scientists don't consider them as breaking down.

Do the 92 naturally occurring elements break down more slowly? And what do we mean by "breaking down"?

Of the 92 naturally occurring elements, only about 26 are found in living things. Of these 26 elements, 6 of them make up nearly the entire weight of all living things. The other 20 elements are necessary for life, but they occur in such minute quantities that they're referred to as "trace" elements.

The six most common elements found in living things are hydrogen, oxygen, carbon, nitrogen, phosphorus, and calcium.

Bio Buzz

An **element** is any substance that cannot be separated into different substances except by nuclear reaction or radioactive decay.

Hydrogen, a gas, is the most abundant element in the universe. Hydrogen gas is so light that it manages to escape Earth's gravity at the upper edges of the atmosphere and then disappear into outer space. Even though it's a gas, about 10 percent of the total weight of plants and animals is hydrogen.

Oxygen is also a gas, but because it's about 16 times heavier than hydrogen, it remains in our atmosphere. Oxygen accounts for about 63 percent of a typical animal's weight, and about 77 percent of a typical plant's weight.

Carbon is an element that's familiar to almost everyone. Those charcoal briquettes that you grill with? Your burnt breakfast toast? The graphite in your pencil? The diamond in your nose stud? That's carbon—all of them. What people generally aren't familiar with is that carbon also exists in their bodies (and not just because you ate that burnt toast). Carbon makes up nearly 19 percent of a typical animal's weight and 12 percent of a typical plant's weight.

Nitrogen, another gas, makes up about 79 percent of the Earth's atmosphere. It's also an important component of the genes and proteins of all living things. About 4 percent of an animal's weight and 1 percent of a plant's total weight is made up of nitrogen.

Phosphorus is not found as a pure element in nature. It's so reactive that it will immediately combine with almost anything, including air. But a certain amount is necessary for life. Phosphorus comprises just less than 1 percent of an animal or plant's total weight.

Calcium is also reactive and is not found in pure form in nature. Hardly any plants have calcium in significant amounts, but for animals, especially mammals, calcium makes up about 2 percent of their total body weight, mostly in the form of bone tissue. Calcium is also necessary for muscle contraction.

So far, we've covered some of the basics of biology, starting with what's alive and what's not, and how the whole field of science and experimentation developed over the centuries. We've also talked a little about matter and the elements, the basic materials of life (and nonlife) on Earth. In the next chapter, we'll be looking at the most current theories on the origins of our solar system and life on Earth. Matter and the elements are critical, but out of all those nonliving things, where did that initial spark of life come from? Let's find out.

> **The Least You Need to Know**

➤ Biology is the study of living things.

➤ The six characteristics of being alive are movement, a high level of organization, homeostasis, energy utilization, reproduction, and growth.

➤ The origins of science are found in curiosity and experimentation. They still move science forward.

➤ All known scientific "facts" have been proven through the scientific method. The scientific method involves challenging your idea through a series of rigorous tests, which will prove the idea's validity.

➤ Everything in the universe, alive or not, is composed of matter. All matter is composed of 106 known elements. Six of these elements (hydrogen, oxygen, carbon, nitrogen, phosphorus, and calcium) account for nearly the entire body weight of all living things.

Chapter 2

In the Beginning

In This Chapter

➤ The origins of life

➤ Creation myths: fact or fiction?

➤ Spontaneous generation

➤ In the primordial soup

➤ What life is made of: the soup ingredients

Scientists (and now you) know that anything that moves, grows, has a high degree of organization, likes constant conditions, uses energy, and reproduces is alive. But where did that first spark of life come from, the one that grew and reproduced?

You're not the only person who doesn't know the answer to that question. Native peoples, priests, scientists, and common folk throughout history have all sought to find, or explain, the origins of life. Lots of answers have emerged, some in the form of creation myths, others in Nobel Prize-winning theories. As you might expect, there is no one correct answer, nor any simple explanation. But over the years, scientists have developed and tested several hypotheses and offer excellent explanations. That's what we'll explore in this chapter.

The "Big Bang" Theory

The origins of life have puzzled people since the first humans appeared on Earth. Life has its origin in the formation of the universe and, specifically, our own solar system. But putting all the evidence together is difficult.

Early native peoples had their own explanations for how the Earth and the living things upon it came to be. Religions from Catholicism to Hinduism have contributed their own theories about the formation of the universe and the development of life.

The scientific community now accepts the *"Big Bang" Theory* of the origin of life. This theory goes something like this: A long time ago ("in a galaxy far, far away …") the explosions of dying stars ripped through space. Scientists think that this giant explosion occurred about 15 billion years ago. After the explosion died down, a thick cloud of dust and gas was left behind. And this was no small cloud—it extended trillions of miles through space.

Within the cloud were carbon and other essential elements that may have been created by the Big Bang or that existed before the explosion occurred. Either way, the raw material needed for the creation of Earth was littered with carbon and other elements.

By about 4.6 billion years ago, the cloud had flattened out into a hot, dense, slowly rotating disc. Within this giant disc of raw material, concentrations of elements and energy developed. Our own sun was one of these "concentrations," and it was right in the middle. For the next 10 billion years, thermonuclear reactions within the concentration of materials kept the sun's development going until it finally became the sun as we see it today. Farther out from the center of the disc, the planets were forming from other concentrations of materials.

At this stage, everything in our solar system was incredibly hot—so hot, in fact, that none of the planets had a stable surface. Everything was oozing magma and chemical reactions. As the planets began to cool, however, a thin crust formed over the slowly revolving mass we know as Earth.

There was no *atmosphere* on Earth at this time. Earth's surface was hot and hard, and meteors ceaselessly bombarded it. Volcanic eruptions occurred constantly on all parts of the globe, and these eruptions are what produced the Earth's first atmosphere. The volcanic eruptions occurred because gases had been trapped beneath the surface when the crust formed. The building pressure forced the gases, along with the molten interior, through the crust. These gases drifted skyward and stayed there, forming the first atmosphere.

Earth's first atmosphere contained virtually no oxygen. An atmosphere without oxygen is called a *reducing atmosphere*. Coincidentally enough, a reducing atmosphere favors the development of life. That's because oxygen disrupts the structure of most things exposed to it: nucleotides, amino acids, and the crumpled bumper on your car. You name it, and oxygen will disrupt it. Without oxygen in the atmosphere, things were able to develop without being disrupted out of existence.

From our standpoint, as creatures that require oxygen to survive, it may seem strange that life can't develop if oxygen is present. That's strange, but true. Oxygen is so potent that it can literally change molecules that come into contact with it. If oxygen had been present when life was forming, life as we know it would never have gotten off the ground. It would have been *oxidized*.

But some oxygen was present during the formation of Earth. Scientists know that the volcanic eruptions released water vapor along with other gases—and, as you remember, water contains oxygen. But the Earth was still so phenomenally hot that the water evaporated the second it hit the air or the planet's surface.

Gradually, however, the crust started to cool down and the water vapor stuck around. This cooling of the planet's surface caused some of the gases in the atmosphere to condense, including oxygen and hydrogen. Torrential rains started falling on the Earth, and they continued to fall (complete with thunder and lightning) for millions of years. The rains eroded the rocks and leached all the mineral salts from them. The water collected in great depressions in the Earth's crust and formed the Earth's first oceans.

At this point, all the "ifs" started coming together. If the Earth had been a smaller planet, it wouldn't have had the gravity to hold on to its atmosphere; the atmosphere would have floated off into space. If the Earth's orbit had been closer to the sun, everything on

Bio Buzz

The **atmosphere** is all the air surrounding Earth. Today, the atmosphere consists mostly of nitrogen, hydrogen, and oxygen.

Bio Buzz

An atmosphere without oxygen, or with very little oxygen, is called a **reducing atmosphere.**

Bio Buzz

Oxidation is the process in which a molecule unites with oxygen, as in burning or rusting. Oxidation alters and changes things. (Some people might even say it destroys things.)

19

Think About It!

If you're looking for interesting reading on the origins of the universe, pick up any book by Carl Sagan. Not only was Sagan a gifted scientist and an excellent writer, but he always kept his sense of wonder and astonishment fully intact. His PBS series, *Cosmos*, is available on video is definitely worth watching.

Weird Science

Any discussion on the origins of life is usually fraught with controversy. Did all the "ifs" come together because of a divine plan, or was it just the Earth's lucky day? This quandary has existed as long as man has been seeking the answers.

the planet would have been baked out of existence. If the Earth's orbit had been farther away from the sun, everything would have been frozen out of existence. But because of where the Earth ended up in relation to the sun, everything was just right—just right for the development of life.

Because of the Earth's size and distance from the sun, the planet was able to retain water on its surface. Without liquid water, life (as we know it) would never have formed. The elements necessary for life came together with the Big Bang and the planet's formation. But water was the medium that allowed everything to function.

The Spontaneous Generation Theory

Myths were only one attempt to explain how life came to be. Early scientists were curious about all kinds of phenomena: Did fish come from mud? Fish appeared in ponds that had been completely dry and devoid of life the previous season. Did flies come from rotting, dead meat? They were always swarming around it. It seemed to some scientists that life could indeed come from nonliving or dead matter, that life could arise by a process called *spontaneous generation*.

Redi's Experiment

An Italian scientist, Francesco Redi (1626–1697), was the first to really test the spontaneous generation theory. He observed that squirming white maggots turned into hard oval cases from which flies emerged. He also observed that the maggots seemed to appear in places where adult flies had previously been. These observations led him to question the commonly held theory that rotting meat spontaneously generated flies.

In 1668, Redi set up an experiment to test the spontaneous generation theory (at least as it pertained to flies and rotting meat). The control group consisted of open jars with meat in them. The experimental group consisted of jars with meat in them, but the mouths of the jars were covered with mesh. The mesh allowed air in, but not insects—and especially not flies.

After a period of time, the meat in both jars rotted, and maggots swarmed over the meat in the open jars. But the mesh-covered jars remained entirely free of maggots, even though flies buzzed desperately around the jars and tried to reach the meat.

The results of Redi's experiment convinced most people that flies did not spontaneously generate from rotten meat. Instead, they were born from eggs laid by other flies.

Spallanzani's Experiment

Redi may have put the idea of spontaneously generating flies to rest, but questions remained about newly discovered microorganisms, seen through a wonderful new invention called the microscope. This amazing invention revealed that the world was teeming with tiny organisms. Scientists decided that because the organisms were so simple and so numerous, they must spontaneously arise from some "vital force" found in the air.

Lazzaro Spallanzani, an Italian scientist (1729–1799), hypothesized that the microorganisms spontaneously generated from other microorganisms, and he decided to try to prove it with an experiment using boiled broth. His control group was a flask of boiled broth that was left open. His experimental group was a flask of boiled broth sealed shut with wax.

Spallanzani's sealed flask remained clear; the open one became cloudy. With nothing but dead microorganisms in his sealed flask, he apparently abandoned the idea of spontaneous generation and concluded that microorganisms came from the air.

For his troubles, Spallanzani's fellow eighteenth-century scientists charged that he'd done it all wrong. By boiling the broth, they claimed that he'd destroyed the "vital force" in the air of the sealed flask. *That's* why the microorganisms couldn't regenerate. (And you thought you had it tough …. Try being a scientist in the 1700s, or any century, for that matter.) As well meaning as Spallanzani's colleagues were, their theory was just as erroneous as was his—and it would take a French scientist about 100 years later to get it right.

Bio Buzz

A **myth** presents itself as a factual account of a true experience. Unlike a fairy tale, which occurs in the time of human experience ("once upon a time"), a myth goes back beyond anything that ever was. The narrative voice of a myth brooks no insubordination. Even the word *myth* comes from the Greek word *mythos*, which means *word*, in the sense of being the final authority on a subject.

Think About It!

To learn more about creation myths, check out a copy of Virginia Hamilton's book *In the Beginning* (Harcourt, Brace, Jovanovich, 1988).

Bio Buzz

Spontaneous generation is the theory that life arises from nonliving matter.

Think About It!

Science definitely carries across the ages. Having trouble seeing the relevance of work carried out in the 1800s? Walk into any grocery store and cruise through the dairy case of the bottled juice aisle. Pasteur's work with spontaneous generation makes it all possible (at least possible without you getting botulism).

Pasteur's Experiment

By the mid-1800s, the spontaneous generation controversy was so fierce that the Paris Academy of Science offered a prize to anyone who could answer the question once and for all. Many people entered the contest, but the winner turned out to be a French scientist, Louis Pasteur (1822–1895).

Pasteur thought that although microorganisms might live in the air, the air didn't actually bring them to life or generate them. For his experiment, Pasteur made a flask with a curved neck and then boiled broth in it. Broth that was boiled in the flask remained clear and uncontaminated for more than a year, even though it was exposed to air. That's because the curved neck trapped the microorganisms from the air, preventing them from reaching the broth. As soon as Pasteur removed the curved neck of the flask, the broth clouded up within one day. Because of these experiments (he did more than one), Pasteur concluded that "productions of infusions [liquids contaminated with microorganisms] previously heated, have no other origin than the solid particles which the air always transports." Microorganisms were certainly in the air, but if you killed them all off in liquid, life couldn't spontaneously generate.

With Pasteur's experiment, the theory of spontaneous generation died a sudden death. Biogenesis, the theory that only living things could generate new life, became the cornerstone of biology and life science.

Pasteur's experiment culminated centuries of attempts to discover the origin of life. From myths with no discernible origin, to experiments with microscopes and glass flasks, science had been moving forward, seeking the answers. Pasteur's experiment didn't answer all the questions (in fact, it answered precious few), but it did eliminate the theory of spontaneous generation—for the time being, and at least as far as microorganisms were concerned.

It was time to move on to the next hypothesis.

In the Primordial Soup

The Earth's oceans and it's oldest rocks formed about 3.9 billion years ago. By that time, the planet had cooled and a crust had formed over the molten ball of rock and gas. Meteors no longer bombarded its surface like wayward missiles. Most of the volcanoes had tapered off. The atmosphere had condensed, and torrential rains fell until the majority of the Earth's surface was covered with water. As old as this history is, it doesn't tell us anything about the Earth's initial formation.

It also tells us nothing about how life began around 3.5 billion years ago, or about 4 billion years after this initial formation. With only rocks, gas, thunderstorms, and volcanoes present, life appears to have spontaneously generated. For 3.9 billion years, there was no life; then for 3.5 billion years, there was life. What happened?

After wading through all other hypotheses and all the experiments using rotting meat and clouded broth, scientists figured out that four things must occur together in order for life to develop:

1. Simple organic compounds (such as amino acids) must be present.

2. Complex organic compounds (such as proteins) must be present.

3. Both the simple and the complex compounds must be concentrated and enclosed.

4. The chemical reactions involved in growth, metabolism, and reproduction must be linked.

For whatever reason, all four of these things managed to occur, simultaneously, on Earth about 3.5 billion years ago. Let's take a good look at each factor.

Bio Buzz

Compounds consist of atoms of two or more elements that are joined together by a chemical bond. Scientists divide compounds into two categories: **organic** (which contain carbon and are usually derived from living things) and **inorganic** (which are derived from nonliving things).

Forming a Simple Organic Compound

We now know that carbon was present at the time of Earth's formation, and scientists think that the other elements needed for life—hydrogen, nitrogen, and oxygen—eventually showed up as gases in the atmosphere. But how these elements actually managed to form a simple organic compound is a true scientific mystery.

During the 1920s, a Russian scientist named Alexander Oparin (1894–1980) developed a theory to explain the initial formation of simple *organic* compounds from *inorganic* matter. Oparin hypothesized that the Earth's early atmosphere contained ammonia, hydrogen gas, water vapor, and *compounds* consisting of hydrogen and carbon (such as methane).

When temperatures were still well above the boiling point of water, these gases could have formed simple organic compounds, such as amino acids. Oparin speculated that when the Earth cooled and the water vapor condensed into rainfall, these simple organic compounds would have collected in the oceans that covered the Earth. He also speculated that, over time, these compounds would have mingled and entered into complex chemical reactions. And, voila! You'd have the beginnings of life.

The only fly in Oparin's particular ointment was that he didn't perform any experiments to test his hypothesis. He formed his theory by studying the writings and experiment results of hundreds of other scientists. But he never verified his hypothesis through experimentation.

Miller and Urey's Soup

In 1953, two American scientists, Stanley Miller and Harold Urey, decided to conduct an experiment using Oparin's hypothesis as a starting point.

They made a chamber that contained all the gases that Oparin had assumed were present in Earth's early atmosphere. Then, as the gases circulated in the chamber, Miller and Urey repeatedly zapped them with electrical sparks to simulate lightning and volcanic activity. (Lightning, intense heat, and even shock waves can provide enough energy to drive a chemical reaction.)

This time it really was "Voila!" In less than a week, the raw materials of life assembled themselves in the chamber-like magic building blocks. Where there had been only chaos and nonliving things, the ordered structures of amino acids—simple organic compounds—were now present in the chamber. It wasn't life itself, but it was the beginning of life. Spontaneous generation was alive and well once again. Miller and Urey's creation was referred to as "primordial soup"—"primordial" because it went to the very origin of life (*primordial* means "primitive, fundamental, or original"—essentially, the very beginning), and "soup" because when everything mixed with the water vapor and condensed, it was a liquid mixture of many things.

Miller and Urey placed gases of ammonia, hydrogen, water vapor, and methane in a sealed chamber. By repeatedly bombarding the chamber with electrical sparks, they were able to reproduce the beginnings of life on Earth.

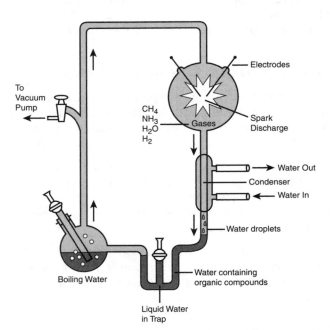

Soup Ingredients

Since the 1950s, scientists have duplicated Miller and Urey's experiment many times. These experiments have produced a variety of simple organic compounds, including various amino acids (the foundations of proteins) and the nucleotides in DNA. DNA is a critical component of life; life can't start or survive without DNA. DNA is a complex molecule that's responsible for the production of new cells and the passing of traits from generation to generation. Once alive, animals need protein to exist, and DNA is also responsible for protein synthesis. The results suggest that the nonliving conditions on Earth may very well have been able to generate life.

But the primordial soup experiment produced only simple organic compounds. What about the other ingredients necessary for life?

Forming Complex Organic Compounds Out of "Nothing"

How did the simple organic compounds form into the complex organic compounds necessary for life?

Like all great mysteries of life, this one is still not totally solved. But we do know how proteins, complex chains of amino acids, may have formed. Some subsequent experiments suggest that amino acids spontaneously formed chains in Earth's early atmosphere, which means that *both* simple and complex compounds formed at the same time, and *both* gathered together in the newly pooling oceans. Other evidence suggests that amino acids will form chains when they're heated in the absence of oxygen.

Either way, there is fairly strong evidence that simple organic compounds would have been capable of forming into complex organic compounds in the early days of Earth.

Complex Organic Compounds in the Sea

The atmosphere is not the only place life may have materialized. Other scientists believe that the pressure, intense heat, and steady supply of high-energy

Think about It!

The formation of organic compounds, and biochemistry in general, is a fascinating subject. To learn more, check out a copy of P. Ritter's book *Biochemistry: A Foundation*, (Brooks/Cole, 1996).

Bio Buzz

Ribonucleic acid, otherwise known as **RNA,** is one of two complex molecules around which all of life is organized. Deoxyribonucleic acid, otherwise known as **DNA,** is the other complex molecule. Both are a type of nucleic acid, or nucleotide. None of the simple and complex organic compounds in the world will form life unless DNA and RNA are present.

inorganic particles found around sea vents on the ocean floor could also have produced life out of nonliving things. Sea vents are openings in the Earth's crust that have been present since the Earth was first formed.

In 1977, a group of scientists in a deep-sea submersible vehicle off the Galapagos Islands stumbled onto a thriving community of creatures clustered around a sea vent. Geysers of water hotter than the core of a nuclear reactor erupted from the sea vent. And there, gathered around the vent, amidst all the heat and inorganic particles, were these creatures.

Further observation and experimentation showed that the communities were thriving due to the presence of an archaebacteria that took in toxic, hot, hydrogen sulfide and used its energy to produce complex organic compounds. Archaebacteria are very simple, microscopic organisms that are able to thrive in habitats that are hostile to all other life forms.

Based on these Galapagos Islands findings, many scientists figure that if sea vents in the ocean's floor can produce life, then the chemical conditions necessary for the formation of life have existed since the Earth's formation. By duplicating sea vent conditions, scientists have been able to produce not only amino acids, but short stretches of *RNA*, a complex molecule related to *DNA* and just as essential to life.

Bio Buzz

A **coacervate** is an irregularly shaped droplet, formed by a group of molecules of different types. A **microsphere** is a spherically shaped droplet that is typically formed by a group of molecules of the same type.

Bio Buzz

A molecule of DNA is formed when two long strands of nucleotides intertwine about each other, forming a shape similar to a spiral staircase. This unique pattern arrangement is called a **double helix**.

Corralling the Compounds

Although not able to produce life, nonliving conditions on early Earth were able to produce the compounds necessary for life, as long as one other element was present. These compounds had to be enclosed within some kind of membrane that would protect them from the environment; otherwise, they would be destroyed before the replicating and chemical reaction stages. The concentration and enclosure of these compounds was the next step on the road to life.

Lab experiments show that complex compounds will clump together on their own to form two types of microscopic droplets, the irregularly shaped *coacervate* and the spherically shaped *microsphere*. Within the droplet, different types of molecules cluster in different areas—again, like a living cell. (Remember, a living cell is "life," so it will have all the components we're talking about here: simple and complex organic compounds that are concentrated and enclosed, as well as the linkage of chemical reactions, such as metabolism and growth.) Also, if you put these things in water, different reactions start taking place. The coacervates grow and the microspheres bud. Under a microscope, both coacervates and microspheres look so much like living cells that scientists have sometimes mistaken them for a new type of bacteria.

Scientists have speculated that somewhere along the line, a coacervate or microsphere enclosed some nucleic acid within its membrane, thus allowing it to carry out the replication process necessary for life to begin. It did so, scientists postulate, when a complex compound and a nucleotide became wet, which caused them to intertwine. After they dried, the complex compound had enclosed the DNA.

Weird Science

A living thing's DNA pattern is unique unto itself, particularly if that living thing is a person. DNA offers such a unique blueprint of a living organism that it is now widely used to pinpoint criminal actions. (Remember the O.J. Simpson trial?) A person's DNA structure identifies him or her better than any fingerprint ever could.

What's So Important About DNA?

For life as we know it to exist, DNA must be present. DNA is the cornerstone of all living things, providing as it does the genetic code for the synthesis of proteins. That is why scientists continue to explore how DNA could have materialized from nonliving substances. If they figure out how, they'll have figured out how life originated on our planet.

In 1953, James Watson and Francis Crick were the first scientists to actually identify and describe the *double helix* structure that is DNA. They determined how many strands of nucleotides made up a molecule of DNA, how those strands bonded together, and how DNA worked. It's important to remember, however, that the work of other scientists had already revealed a great deal about DNA, including

Bio Buzz

Metabolize means to utilize energy. A plant obtains energy from the sun and metabolizes it into carbohydrates, proteins, and fats. An animal eats a plant and metabolizes its energy into other functions.

its basic chemical structure and molecular shape, before Watson and Crick made their discovery. Furthermore, Watson and Crick were able to do their work only because of technological advances—such as X-ray crystallography and a model-building technique developed by Linus Pauling—gave them the right tool.

Unlike Galileo, Spallanzani, and other scientists, Watson and Crick actually received kudos for their efforts. In 1962, they were jointly awarded the Nobel Prize.

Go Forth and Multiply!

Coacervates and microspheres look "real," but they're not. They lack the complexity of living cells, and they can neither maintain stable growth nor reproduce. They also can't *metabolize*. But they're pretty close to the real thing.

Scientists speculate that over time, a few coacervates and microspheres enclosed DNA and managed to develop these traits and abilities. Having DNA allowed coacervates and microspheres to metabolize and reproduce, thus allowing a nonliving thing to be able to cross over to the land of the living.

Laid out on paper, it all looks simple enough. But although the hypotheses and experiments look convincing (they *are* convincing), scientists have yet been able to make DNA in a laboratory setting. They haven't been able to create DNA from nonliving materials. But they've come very close, and for now, that's as good an answer as anyone is going to get to the mystery of life.

We've covered a lot of things in this chapter: the origins of our solar system and how life emerged from nonliving matter during the early days of Earth. We also discussed how scientists have replicated some early life forms under laboratory conditions. In the next chapter, we'll talk about the development of these early life forms and how scientists study them.

The Least You Need to Know

➤ Life originated from the nonliving matter that formed our solar system.

➤ The conditions necessary for life started with the Big Bang, the explosion that commenced our solar system's formation. The Big Bang also produced a scattering of carbon, hydrogen, nitrogen, oxygen, and other elements that are essential for the development of life.

➤ Life first appeared on Earth between 3.9 billion and 3.5 billion years ago, about the same time the oceans formed.

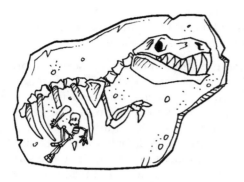

Chapter 3

History of Life on Earth

In This Chapter

➤ The big picture: macroevolution

➤ The fossils never lie

➤ Measuring geologic and evolutionary time

➤ Mass extinctions: the dinosaurs

As discussed in Chapter 2, "In the Beginning," life first materialized on Earth somewhere between 3.9 billion and 3.5 billion years ago. Considering that the origins of Earth stretch back at least 15 billion years, life was a long time coming. And it hasn't been here very long in comparison to the existence of the planet itself.

Earth's first life forms didn't look anything like the complex animals and plants we know today. Earth's first inhabitants were simple cell structures, but they were no less alive than human beings are today. Alive is alive.

Scientists have spent hundreds of years researching how life reached the stage it's at today. It has taken a while to decipher the message, but scientists have got much of it. Many of these previous life forms left clues about their existence. There is a pattern and a record—you just have to know how to look for it!

In this chapter, we'll be talking about how life forms evolved simultaneously around the globe, and the different types of things scientists look for when they're examining the evolution of life. We'll also talk a little bit about the flip side of life, specifically, the death of the dinosaurs.

Macroevolution

The history of life on Earth spans nearly 4 billion years, and it's a very rich history indeed. It's the story of how life originated, existed, became extinct, stayed in one place, or radiated into new environments—in short, its about *evolution*.

Bio Buzz

Evolution refers to the process by which a species or an environment changes over a period of time, evolving into different forms. Through evolution, the character of any given population changes with successive generations. The word *evolution* means "unrolling or unfolding." I understand what you're saying here (the ocean evolves, the prairies evolve), but in a biology book, the term refers only to species. For simplicity, take out "or an environment" from the main definition because it clouds the main issue.

Bio Buzz

Mutation is the process by which a molecule of DNA alters its original pattern, thus altering the life form.

Macroevolution is evolution on a large scale, involving large-scale patterns, rates of change, and trends among whole groups of species. Microevolution, on the other hand, examines the pattern of change through time of a single species. Studying the microevolution of a single species helps scientists develop theories about macroevolution, since the same principles of pattern change and modification to existing life forms apply. As discussed, the first life form on Earth may have spontaneously generated from nonliving materials. Once this first living thing existed, however, scientists believe that all other life forms evolved from it. New species don't just appear out of thin air.

The most obvious question here is, "Why not? New species appeared out of thin air to start with!" (Or, more appropriately, thick, toxic air.) But think of it this way: The very specific conditions that allowed life to generate spontaneously ceased to exist because life itself fundamentally changed the environment. Right from the start, the appearance of life halted the spontaneous generation of more life. The chemical reactions produced by the life forms generated different gases, and Earth's atmosphere changed from a reducing to oxidizing one. The life forms needed the oxygen to continue to exist, but the oxygen eliminated the possibility that new life could spontaneously develop.

From then on, and even now, new species emerge through *mutation*, natural selection, or other genetic processes, usually in populations that are reproductively isolated. But they always develop from previously existing life forms.

When scientists look for patterns, they look for both those traits retained by a species and those traits modified between species. Taken together, these patterns can indicate what common threads exist between species and groups of species.

If the theory of evolution is true, then all species would be connected all the way back to the origins of life. The study of macroevolution reveals common threads like the similarities of body forms and similar structural patterns within major groups.

A branch of science that further explores these evolutionary similarities is comparative morphology. The work involves detailed analyses of embryos at different stages of development, as well as detailed analyses of the organism's adult form. The theory is that the shape of an organism's body is the result of the evolutionary process.

Many different organisms look virtually identical to one another at certain stages of their embryonic development. Frequently, these similarities indicate an evolutionary relationship. For example, certain structures emerge in all vertebrates (animal species that have backbones) at corresponding stages of development, and they emerge in the whole class of vertebrates, not just mammals, birds, or reptiles.

Think About It!

Evolution is happening all around us, right now. In fact, it's happening much faster than scientists ever thought. For a great read on the subject, pick up a copy of *The Beak of the Finch*, by Jonathan Weiner. It tells the story of two Princeton University scientists and the research they're doing with finches on the Galapagos Islands. It's interesting, thought-provoking, and funny.

For instance, nearly all land-dwelling vertebrates have a five-toed limb. This limb was the evolutionary departure point for the development of wings and flippers in air and sea creatures. All types of wings have the same component parts, as do the flippers on porpoises and other sea mammals. The existence of these homologous structures indicates that these different species evolved from common ancestor.

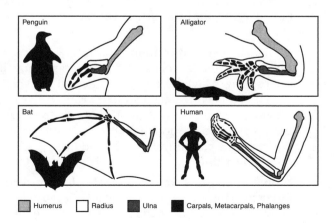

The forelimbs of humans, bats, alligators, and penguins all have the same structure in the embryonic stage.

A lot of evidence for the theory of macroevolution comes from the fossil record. Through fossils, scientists are able to compare hundreds of examples of microevolutionary development in order to develop a planetary evolutionary picture.

The Fossil Record

Fossils are the most overt evidence we have that life existed before mankind walked on the Earth.

About 500 years ago, Michelangelo was completing his painting of the Great Deluge on the ceiling of the Sistine Chapel in Rome, Italy. In it, he depicted the great biblical flood that wiped out all life on Earth except those creatures rescued by Noah. At the same time, another Italian artist, Leonardo da Vinci, was pondering the presence of completely intact seashells embedded in the rocks of the towering mountains of northern Italy. The shells were miles from any ocean. How did they get there?

Bio Buzz

A **fossil** is the hardened remains of a plant or animal from a previous geological age, which has been preserved in the Earth's crust.

Think About It!

For an interesting read, check out J. Horgan's article, "Trends in Evolution: In the Beginning" on pages 116–125 in the February, 1991 issue of *Scientific American* 264(2): 116–125.

At the time, the traditional view was that the shells had been washed up on the mountain peaks during the Great Deluge. But da Vinci saw flaws in the traditional view. How could even a great flood wash the shells miles inland and thousands of feet up the mountainside? Many of the shells were so thin and fragile that they never would have survived being dashed against a mountainside. Leonardo da Vinci also perceived that the rocks of the mountains were stratified, or laid down in layers like a cake. Some layers contained only one type of shell, others contained a mix, and still others contained no shells at all.

So, da Vinci started thinking about how Italy's great rivers deposited layers of mud when they flooded each spring. Maybe the mountains, and the shell layers, had been formed through years and years of flooding and silt deposition. If that was the case, then maybe he was looking at the remains of ancient marine communities that had been gradually buried over time.

Leonardo da Vinci was no fool. He didn't run about discussing his radical idea, probably because he knew he'd be imprisoned, or worse. But the idea took shape and was later more fully explored by other scientists.

Paleontology Develops

By the 1700s, the field of paleontology started to flourish, primarily because so many fossils were being found. Amateur paleontologists were curious about the fossils and what they represented. The fossil forms and the stories they told were so diverse. What were these things?

Fossils are formed in a variety of ways. The most common fossils are found in sedimentary rock and are formed from the hard parts of organisms, such as bones, teeth, shells, and the woody stems of plants. For a fossil of a body part to form, silt has to cover the organism's body almost immediately upon death, before decomposition can occur. Sometimes minerals replace the original body parts, molecule by molecule. When this type of fossilization occurs, the microscopic structure of the organism is preserved.

Sometimes the fossils are of a footprint, a track, or some other impression. An imprint is a type of fossil where only a film of carbon remains after everything else has decayed. Plants typically produce an imprint fossil, with only a black or brown film on the rock recording the details of the leaf. A mold is a type of fossil where an impression, such as a footprint or track of an organism, has survived.

In the 1700s, all these fossilized plants, bones, shells, and impressions of tracks, burrows, and footprints were acknowledged to be evidence of life in the past. But they were still being interpreted within the traditional religious viewpoint: The fossils were the remains of creatures that had perished during the Great Deluge. For instance, when a sixteenth-century Swiss naturalist found a large skeleton, he excitedly announced that he had found the remains of a man that had perished during the Great Deluge! Unfortunately, the "man" was actually the remains of a giant salamander. Except for a few freethinkers (like da Vinci), who didn't broadcast their ideas, all fossils were explained in relation to the Bible.

Uncovering the Layers of Life

It wasn't until the 1800s that scientists began to publicly question the traditional religious viewpoint. More and more fossils were being unearthed as quarries and mines were dug to support the building of new roads, cathedrals, and homes, and it was getting hard to explain the existence of these fossils in biblical terms. In addition, global exploration was opening up whole new areas of natural history and paleontology. Asia, Africa, and the New World were found to have hundreds of exotic animals and plants, all of them unknown in Europe. If all living things had been created at the same place at the same time, and then later released from Noah's Ark, why were the distribution patterns so different?

All these social and scientific changes opened up new avenues of interpretation for paleontologists. It didn't take them long to prove that da Vinci had been right: Fossils did occur in layers, and the layers appeared to represent some sort of timeline, perhaps with the deepest layers being the oldest and the highest layers being the newest. In addition, the layers occurred simultaneously over enormous distances.

Think About It!

Want to learn more about early times on the earth? Read E. Dobb's article, "Hot Times in the Cretaceous," in the February, 1992 issue of *Discover* 13: 11–13.

Bio Buzz

The doctrine that creation had occurred several times, only to be wiped out by a flood each time, was referred to as **catastrophism.** While taking fossil layers and differences into account, catastrophism didn't really challenge the prevailing religious viewpoint.

Changes in scientific thinking were afoot, but they didn't come quickly. One French anatomist, George Cuvier, spent more than 25 years comparing fossils with living organisms. He perceived that the fossil record changed abruptly at several different layers. He hypothesized that there had been not one, but six different creations, each of them wiped out by an enormous flood. His attempt to explain the fossils became known as *catastrophism.*

An English geologist, James Hutton, was one of the first people to tie the formation of fossils into the formation of the Earth when he published his book, *Theory of the Earth,* in 1788. Hutton proposed that the same principles that were shaping the Earth today—weathering and erosion, deposition of sediment, and volcanic activity—had also shaped the Earth over time. For all of these changes to have occurred, the Earth had to be very, very old.

Hutton was the first to see that natural processes could shape both geological and biological features. He was also the first to link biology and geology with the concept of deep time. When conducting his geological research, Hutton was overwhelmed at how old the Earth apparently was. He really hit the nail on the head when he announced, "The result … of our present inquiry is that we find no vestige of a beginning, no prospect of an end."

The Incomplete Fossil Record

As many areas of scientific thinking came together, scientists began looking at fossils differently. They realized that the fossils really did represent a record of the development of life on Earth and, as such, could answer important questions about evolution. However, there is no question that the fossil record is an incomplete one. Not all organisms died under the right conditions to become fossils, for instance, nor could soft-bodied organisms, like jellyfish, become fossils.

Furthermore, not every geographic area is represented in the fossil record. While swamps, flood plains, sea floors, caves, and tar pits are natural fossil traps, hills and mountains subject to erosion have a poor fossil record. The Grand Canyon, for instance, has some of the most magnificent layering on the planet, but it has been subject to such erosion that any fossils that have emerged have been swept away.

Finally, the fossil record is incomplete in terms of the quality of the specimens represented. Scientists love it when they find an entire skeleton, completely intact and perfectly preserved. But these kinds of fossil finds are rare. More often, the original organism was broken, incomplete, or crushed before it ever became a fossil. Or, it was complete but was damaged by time and weather forces after fossilization.

Because of these insufficiencies, the Earth's fossil record doesn't answer some important questions about life on Earth. We'll never know, for instance, what type of organisms lived in areas with rapid erosion. But scientists have identified about 250,000 fossils, mostly dating from the last 600 million years.

Try It Yourself

Real fossils take forever (virtually) to occur. But you can duplicate the molding and imprint process in less than a half-hour.

To understand the process of fossilization, try this simple experiment. Mix up some plaster of paris (available in most hardware and home improvement stores) with enough water to make a goopy mess (you can also follow the directions on the package). Now comes the fun part. Pour it into the mold of a foot, hoof, or paw print. (If you live in a rural area, finding a print is probably not a problem. In urban areas, look for dog prints in mud, or use an impression of your own hand in sand.)

The plaster should set up in less than a half-hour, at which point you just lift it off the ground. You should have a cast of the print. You can also make other "fossils" by sticking a shell, a bit of apple, or any other object into a container of plaster. Let it harden, and then chip away the plaster.

Among the 250,000 fossils we do have, some lineages are quite extensive. For example, when it comes to a shallow seas environment in which sediments are steadily deposited, scientists have a nearly complete record. The number of shelled fossils is pretty close to the number of species still living in these environments. In a shallow, sedimentary sea, all the conditions for fossilization came together. The organisms had or still have hard shells, layers form on a consistent basis, and the layers are often undisturbed. Nearly 90 percent of all the known fossils are of marine organisms.

For us, the best known fossils and evidence of early life come from things we can identify: woody plants, seashells, and vertebrates. We recognize these fossils because

Bio Buzz

A **microfossil** is a fossilized cell, or chain of cells, that can be seen only with a microscope. Scientists believe that microfossils offer evidence of the beginning of life on Earth: a simple, spontaneously generated life form. These are the organisms from whence everything else sprang.

the organisms are still on Earth; we live with these things every day. But within the fossil record, these types of organisms show up pretty far down the line. It's a long leap between the first life forms and a vertebrate. A lot of evolution, both micro and macro, occurred before a vertebrate or a woody plant came on the scene.

The earliest fossils, the ones from which all life dates, are simple-celled creatures that have been found in Australia. And these things are not only primitive in form—they're microscopic. These *microfossils* are the original, real McCoys.

Although these first fossils are ancient, they're probably not the first cells that ever lived. Most likely, these fossilized cells are the descendants of a long line of unpreserved cells.

According to the theory of macroevolution, these are the cells that started the web of life on Earth as we know it today. But how do scientists know this? How do they order and date the fossils they find? What determines a fossil's age and place in the overall record?

Measuring Geologic and Evolutionary Time

When geologists started mapping the Earth's layers in the 1800s, they knew that sedimentary layers were deposited through a gradual "rain" of silt, debris, and the bodies of marine organisms. It was logical to assume that the deepest layers had formed first and that the layers closest to the surface had formed last.

Scientists also noticed that the deepest layers had the fewest life forms, as well as the most simple life forms. The closer a layer was to the surface, the more complex, numerous, and diverse were the life forms contained in it.

Scientists believed they were looking at a timeline of some sort, but it was a timeline that had no absolute point. The relative age of a fossil was determined by its place in the sedimentary layers, but scientists needed a way to actually date the fossils and link them with the geologic time frame.

It wasn't until the 1900s that a way was found to correctly date fossils and geologic layers. Nowadays, the age of a fossil is determined through *radioactive isotopes.*

All living things contain carbon. Carbon 12 is a common and stable isotope of carbon. It has six protons and six neutrons (hence, the "12"), and it doesn't decay. But carbon 14 is an unstable isotope of carbon that has six protons and eight neutrons. Carbon 14 decays into nitrogen 14 and has a *half-life* of 5,700 years.

In Earth's atmosphere, the ratio of carbon 12 to carbon 14 is always constant. Living organisms take in both isotopes equally. When an organism dies, the ratio of carbon 14 to carbon 12 will be the same as the ratio of the atmosphere.

But as time goes by, carbon 14 will break down within the organism's body or within the fossil. By comparing the ratio of carbon 14 to carbon 12 in the fossil, and then comparing that number to the ratio in the atmosphere, scientists can determine the age of the fossil because they know the rate and the pattern.

Scientists use other radioactive isotopes to date older fossils. Potassium 40 has a half-life of 1.28 billion years. Uranium 238 also has a long half-life and is found in sedimentary rock. Scientists use uranium 238 to date the age of cores of silt taken from the ocean or ancient sea beds.

Bio Buzz

Radioactive isotopes have an unstable molecular structure that causes them to decay and form into other elements. The isotopes decay at a known rate. The period of time it takes for one-half of the unstable isotope to decay is called the **half-life** of the isotope.

Mass Extinctions: The Dinosaurs

No discussion of fossils would be complete without mentioning those giant creatures, "the dinosaurs!"

Clearly, dinosaurs once existed on Earth and, for whatever reason, became extinct. Something happened on Earth, and that was it—kaput! No more dinosaurs.

What few people realize (aside from those who study life science—an elite group, which you are now part of), is that the extinction of the dinosaurs was just one of a series of mass extinctions that have occurred.

Remember those abrupt transitions in the sedimentary layers that gave rise to the doctrine of catastrophism? Well, George Cuvier wasn't far off the mark. Each abrupt transition in the fossil record represents a *mass extinction*.

Bio Buzz

A **background extinction** is a normal, single species extinction that can be viewed through the fossil record and sedimentary layers across time. A **mass extinction** refers to an event in which whole major groups of species become extinct at the same time. Mass extinctions are caused by catastrophic global events.

When scientists look at lineages, some species inevitably disappear as local conditions change. Across the fossil record, there is a fairly steady rate of extinction. Scientists call these *background extinctions*. In contrast, a mass extinction is an abrupt rise in the

Think About It!

Think dinosaurs! For a realistic look at these by-gone animals (as well as a good thrill), watch *Jurassic Park,* Steven Spielberg's blockbuster production. Many scientists think that when it comes to dinosaurs, Spielberg got it right.

extinction rate as compared with background extinctions. In the fossil record, a mass extinction indicates a catastrophic, global event in which whole major groups of species are wiped out.

Major groups of species usually can survive a mass extinction event if their numbers are dispersed over different areas of the globe. Groups hardest hit tend to be those with very specialized ways of life. Organisms living in the tropics, for instance, have very small parameters when it comes to temperature ranges. If the environmental temperature rises or falls by just a few degrees, the Earth can lose one or more tropical species.

Luck also plays a large part in surviving a mass extinction event. Having all of the adaptive traits in the world won't help a species survive if an asteroid the size of Texas hits next door to it.

Before any lineage can successfully occupy an environment, there has to be room for it, and mass extinctions clear the way for other species to move into new niches. For instance, the dinosaurs were gigantic, took up enormous amounts of room, consumed huge quantities of food, and terrorized other species. If they had not suffered a mass extinction, Earth might never have had the room for any other mammals. The extinction of the dinosaurs allowed smaller, high-metabolism organisms to move into previously occupied niches.

Within the fossil record and the development of life on Earth, there have been at least five mass extinctions. The last one occurred about 65 million years ago and destroyed all the dinosaurs and many marine organisms.

Scientists have divided the geological and biological history of Earth into 5 major eras, with about 17 different sub-eras. Starting from the beginning, the five major eras are these:

➤ The Archean era (beginning of life to 2,500 million years ago)

➤ The Proterozoic era (2,500 million years ago to 550 million years ago)

➤ The Paleozoic era (550 million years ago to 240 million years ago)

➤ The Mesozoic era (240 million years ago to 65 million years ago)

➤ The Cenozoic era (65 million years ago to the present)

A million years might sound like a long time, but it's a veritable blink on the cosmic time scale.

Think of the eras as taking place on the face of a clock, with life originating at midnight. The Paleozoic era wouldn't even start until 10:04 A.M. The Mesozoic would start at 11:09 A.M., and the Cenozoic would start at 11:47 A.M. In less than 12 hours, we've covered the history of life on Earth.

In studying the fossil record, scientists are always coming up with new information. New techniques and developments in scientific research not only yield new evidence, but also provide better ways to interpret existing evidence. The more scientists research, the more it all seems to go back to that first spark of life.

The Least You Need to Know

➤ The Earth is at least 15 billion years old, and life has existed on Earth for approximately the last 4 billion years.

➤ Evolution is the process by which species change over time. Microevolution is how individual species evolve and change. The study of macroevolution looks for patterns, rates of change, and trends among whole groups of species.

➤ Fossils offer evidence of previous life on Earth. Fossils occur in sedimentary layers in the Earth's crust. The fossil record combines geologic and evolutionary events into one time line.

➤ Geologic and evolutionary time is divided into the Archean, Proterozoic, Paleozoic, Mesozoic, and Cenozoic eras. The delineation between eras is frequently determined by a mass extinction, such as the extinction of the dinosaurs.

Chapter 4

Understanding the Cells: The Building Blocks of Life

> **In This Chapter**
> ➤ Life: Understanding the building blocks of life
> ➤ Exploring different cells
> ➤ Cell division: divide and conquer?
> ➤ Bigger than a breadbox? Smaller than a bee?

While the origin of life may be found in the stars, its continuation is found in cells. Cells are the basic building blocks of life. Scientists believe that they have figured out how the first cells generated from nonliving materials, but what happened after that? It's a big leap from a single cell to a polar bear or an oak tree.

In this chapter, we're going to be talking about the discovery of cells, how all life originates from cells, and the different types of cells that there are.

The Building Blocks of Life

Every living thing on Earth is composed of cells—from a floating phytoplankton, to a peach tree, to an elephant, to you. Cells are the building blocks, where everything starts, yet most are microscopic. A large organism, like an elephant, isn't just a single cell; it's composed of billions of microscopic cells. All these teeny little cells are vital to the functioning of life. For instance, life on Earth, as we know it, depends on single-celled phytoplankton that float in the oceans. That's because these organisms produce nearly all the oxygen on this planet. Without these single-celled organisms, we wouldn't exist. But seven million phytoplankton laid end to end would be less than a quarter-inch long. If most cells are this tiny, how did scientists even discover them?

Surprisingly, astronomer Galileo was the first person to record his observation of cells. Early in the seventeenth century, Galileo arranged two glass lenses together to create a primitive microscope that allowed him to look at an insect in a whole new way. What he saw astounded him. In his notebooks, he described the stunning geometric patterns of the insect's eyes. What he was really seeing, was a grouping of cells (albeit, in a geometric pattern). Of course, Galileo didn't know this. Scientists who came after him are the ones who really verified that cells made up the eyes of insects.

Galileo's observations marked the beginning of the study of the cellular basis of life. In 1665, Robert Hooke, an English scientist who was Curator of Instruments for the Royal Society of England, observed a dried piece of cork under his microscope. He, too, was astonished. In his notebooks he wrote, "I could exceedingly plainly perceive it to be all perforated and porous." He also wrote that it consisted of "a great many little Boxes." Hooke decided to call the "little boxes" cells because they reminded him of the small rooms monks live in.

What Hooke had actually observed were dead plant cells. But he didn't think of them as "dead" because he didn't know that cells could be alive. A Dutch shopkeeper, Anton van Leeuwenhoek, was the first person to observe living cells. And while van Leeuwenhoek might have been a shopkeeper, he was skilled in constructing and using microscopes. He also had amazing eyesight. In 1675, he observed a single bacterium, a type of organism so small that it wouldn't be seen again for another 200 years.

Building a Theory

Hooke and van Leeuwenhoek lived in an age of exploration, not interpretation. They knew what they'd seen, they recorded it, and that was pretty much the end of the matter. It wasn't until the mid-1800s that other scientists organized these observations and hypotheses into a coherent theory.

In 1838, a German botanist, Matthias Schleiden, concluded that all plants were made up of cells. A year later, in 1839, a German zoologist, Theodor Schwann, concluded that all animals were composed of cells. Then, in 1855, a German physician, Rudolph Virchow, determined that cells arise only from other cells during his study of how disease affected living things.

Taken together, the observations of these three scientists became known as the Theory of Cells (or Cell Theory, for short). This theory has three tenets:

1. All living things are composed of one or more cells.

2. Cells are the basic units of an organism's structure and function.

3. Cells come only from existing cells.

The Cell Theory established that cells are not only the smallest living unit, but that they also are responsible for an organism continuing to grow and exist. Life progresses through the growth and division of each tiny cell. Within each cell, events occurred that had a profound effect on every level of biological organization.

Every living thing is composed of cells, but not all cells are alike. Most organisms exhibit enormous diversity in size, shape, function, and organization, and their cells reflect that diversity as well. Take their size, for instance. The cells in the red part of a watermelon, the yolks of bird eggs, and the cells of fish eggs (otherwise known as caviar) are so big you can see them with the naked eye. Giraffes also possess long, thin nerve cells running down their legs that are longer than 2 meters. But most cells are microscopic in size. For instance, if you made a chain of your own red blood cells, 2,000 of them would barely stretch across the width of your thumbnail.

Cells also come in a variety of shapes. Most are roughly spherical or cuboidal. But many plant cells are shaped like triskaidekahedrons (say that one fast, three times), which means that they're roughly round but have 13 flattened sides. When it comes to cells, the variety in shape and size reflects the variety of function.

Cell Structure and Function

Cell complexity varies from simple to intricately complex. But in a few basic areas, all cells are alike.

All cells have a *plasma membrane* that surrounds an inner region. The plasma membrane is the cell's outermost layer. It separates the internal events from the environment so that they can proceed in organized, controlled ways. If you think of a cell as a blown-up balloon, the plasma membrane would be the balloon itself; it contains the inner workings of the cell.

The inner region of the cell is called the *cytoplasm*. The cytoplasm is nearly everything that is enclosed by the plasma membrane. Going back to the balloon analogy, you could say that the air in the balloon represents the cytoplasm.

A cell's cytoplasm is full of all sorts of things, most notably, organelles. An organelle is a membrane-bound sac. A cell might have many organelles in its cytoplasm, or almost none. There are also different types of organelles in the cytoplasm, each one performing a different function for the cell. For instance, some organelles enable the cell to produce proteins, while others help metabolize nutrients.

Within the cytoplasm, the cell's genetic material is concentrated in one area, called the *nucleus*.

Think About It!

Want to know more about cells and how they function? Check out a copy of C. deDuve's book, *A Guided Tour of the Living Cell* (Freeman, 1985). Everything you'll ever want to know, in two short volumes.

Bio Buzz

The **plasma membrane** is the cell's outermost layer. It separates the inner region of the cell from the environment. The inner region of the cell is called the **cytoplasm.** Within the cytoplasm is the **nucleus.** The nucleus encloses the cell's genetic material. In prokaryotic cells, the **nucleoid** contains the genetic material.

Bio Buzz

A **compound** is formed whenever the atoms of two or more elements join together in a chemical bond. An **organic compound** is a compound made up of organic elements, those elements that are needed to form life, such as carbon, hydrogen, or oxygen.

The nucleus is a membrane-bound compartment that houses all the cell's genetic material. The nucleus contains the cell's DNA. Although the nucleus could technically be considered an organelle (because it is indeed a membrane-bound sac), because it's so important, scientists refer to it as the nucleus, not just an organelle. It's hard to say that one part of a cell is more valuable than another because all parts of a cell function together, but a cell's nucleus is definitely important, the way your own brain is.

In recent years, DNA has been the subject of much research. Because the DNA in any cell's nucleus essentially controls how the cell grows, functions, and reproduces, scientists want to unravel its mysteries. If scientists can better understand DNA, they will better understand the process of life itself. Bacteria and cyanophyte cells are the only cells that don't have a true nucleus. But they still possess genetic material. The area that houses their genetic material is called the *nucleoid.*

The "C" Word: Chemistry

You can't talk about cells without talking about chemistry. Cells *are* chemistry. A cell's composition and development is the result of chemical reactions, the very reactions that chemistry—the field of science that deals with the composition and properties of different substances—is all about.

You already know about atoms (the smallest particle of any element) and molecules (the smallest part of an element that can exist freely by itself). If you recall, a *compound* is formed whenever the atoms from two or more elements join together. A compound can be formed from any number of elements. For instance, combining iron and carbon forms steel. But for our purposes, we're going to be concerned about only organic compounds. Any chemical compound that makes up a living thing is referred to as an *organic compound.* Specifically, we're going to discuss carbohydrates, lipids, proteins, and nucleic acids because those are the organic compounds necessary for life to exist.

One type of organic compound is a carbohydrate. Every organism, including all cells, uses carbohydrates as sources of energy. Some carbohydrates even serve as structural material for cells. Carbohydrates are composed of carbon, hydrogen, and oxygen. In a carbohydrate, the ratio of hydrogen atoms to oxygen atoms is 2:1.

Lipids are also composed of carbon, hydrogen, and oxygen (are you noticing a pattern here?), but the ratio of hydrogen atoms to oxygen atoms is much higher. Lipids include such compounds as steroids, waxes, and fats. Lipids that are contained in cells are a source of stored, concentrated energy.

Proteins are immense in size and are very complex. They're among the most complex group of organic compounds. Proteins are made up of units of amino acids. Amino acids contain carbon, hydrogen, oxygen, and nitrogen. Some amino acids contain sulfur and phosphorus atoms, as well as trace elements such as copper or iron. Proteins are the major molecules from which all living things are constructed. One reason proteins can be difficult to understand is because they're an organic compound that's produced in steps. First the organic compounds of amino acids need to form through their own combinations of carbon, hydrogen, oxygen, and nitrogen, and *then* the proteins (which are also an organic compound in their own right) form through combinations of amino acids. You can think of it like a car: First the steel has to be made, and then the car is made from the steel.

The amazing thing about cells is that, through chemical reactions, they are able to take all these elements (oxygen, carbon, and so on) from the Earth, the atmosphere, the food you eat, and wherever they can get them, and assemble them into organic compounds and life itself.

Last, but not least, there are nucleic acids. Just like protein, nucleic acids are large, complex molecules. Nucleic acids are composed of smaller molecules called nucleotides. Each nucleotide contains a carbohydrate molecule, a phosphate group, and a nitrogen-containing molecule. The nitrogen-containing molecule is referred to as a nitrogenous base. (If proteins are a car, then nucleotides are a Lear jet.) All living organisms possess two important types of nucleic acid: deoxyribonucleic acid (DNA) and ribonucleic acid (RNA), the true building blocks of all life. DNA is found primarily within the nucleus of the cell, while RNA can be found in both the nucleus and in the cytoplasm. Given how critical DNA is to a cell and its importance in controlling cellular functions, it's somewhat logical that it would be such a complex molecule or organic compound.

Think About It!

Carbohydrates are some of the most important compounds used in cellular activities. To find out more about how your body uses carbohydrates, read Liz Applegate's article in *Runner's World*, October 1990, page 24, titled "Carbo-hydration."

Think About It!

For a comprehensive overview of the structure and functioning of biological molecules, check out a copy of the October, 1985 issue of *Scientific American*. The entire issue is devoted to the subject.

Less is Less?

Cells are cells, right? Well, sort of. Scientists divide cells into two basic categories: prokaryotic cells (which are very simple, have few organelles, and aren't capable of much complex functioning) and eukaryotic cells (which are complex, have more organelles, and can perform all sorts of functions). You are composed of eukaryotic cells.

Prokaryotic cells are the smallest and, in structural terms, the simplest cells. The first cells on Earth were prokaryotic, meaning that that first spark of life millions of years ago produced a prokaryotic cell. Like everything else in life, cells evolve, too. Today the only types of cells that are prokaryotic are bacteria and cyanophytes. (We'll discuss bacteria in more detail later in this book.) Cyanophytes used to be called blue-green algae because they are blue-green in color and use sunlight to make energy, like plants do. But cyanophytes are extremely similar to bacteria, particularly because algae are eukaryotic organisms. Many scientists now refer to cyanophytes as cyanobacteria, to emphasize that they are more closely related to bacteria than algae.

Prokaryotic cells have a plasma membrane surrounded by a semirigid wall. Secretions from the cell form the outer wall, supporting the cell and giving it shape.

Prokaryotic cells have a smaller volume of cytoplasm, and the cytoplasm is more simplistic than in other cells. The cytoplasm of prokaryotic cells has *ribosomes* (tiny molecular units) dispersed through it, but that's about all.

Besides having a simple structure with virtually no organelles (ribosomes could be considered organelles), the greatest distinguishing factor of a prokaryotic cell is that it doesn't have a nucleus. The DNA of a prokaryotic cell floats freely in the cytoplasm and is irregularly shaped.

Prokaryotic cells don't have much complexity, but they have ways of obtaining nourishment and getting around. Many bacterial cells have pili or flagella. Pili are projections of the cell's plasma membrane that allow for attachment to another bacterium for a transfer of genetic material, or to parasitize another cell. Flagella are whip-like appendages. The word *flagellum* derives from *flagellate*, which means "to whip." The flagellum whips about like the propeller of a boat. This whipping motion enables the cell to move about. Most of a human sperm cell's length comes from its flagellum. Some bacteria have no flagella or pili, others have 1 or 2, and still others might have 100.

Bio Buzz

The word **prokaryotic** means "before the nucleus." The fact that scientists named certain cells in this way implies that they believed that some forms of bacteria existed on Earth before the evolution of cells with a nucleus.

Bio Buzz

A **ribosome** consists of two molecular units, each composed of RNA and protein molecules. In all cells, ribosomes are responsible for making proteins.

Cyanophytes never have pili or flagella. But because they can make their own nourishment from sunlight, they don't need to parasitize other cells or move around.

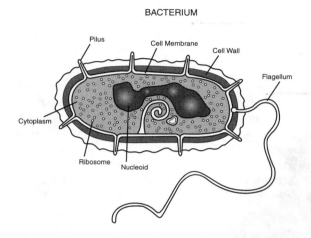

BACTERIUM

All prokaryotic cells possess a cell membrane, cell wall, cytoplasm, and nucleoid. Bacteria cells may also possess pili or flagella.

The Basics of Animal and Plant Eukaryotes

From an evolutionary standpoint, eukaryotic cells are more advanced than prokaryotic cells. Both possess a plasma membrane and cytoplasm, but after that, differences emerge. The cytoplasm of eukaryotic cells is quite complex. Both animals and plants are composed of *eukaryotic cells*.

Cytoplasm is a jelly-like material. It surrounds the nucleus and organelles, and contains water, salts, and organic molecules. The cytoplasm is in constant motion as the organelles and particles jostle and move about. Scientists refer to this constant motion as cytoplasmic streaming.

Bio Buzz

A **eukaryotic cell** has a plasma membrane, cytoplasm, and a nucleus. Within the cytoplasm are **organelles,** membrane–bound regions that enclose specific chemical activities.

Get Those Organelles to Work, Work, Work!

The organelles are where all the action is in the cell, and animal cells have a number of different kinds of these mini-organs, each with their own set of functions.

One type is the endoplasmic reticulum (ER). The ER is a series of membranes that extends throughout the cytoplasm of eukaryotic cells. In some places, the ER is studded with ribosomes, which make the cell's protein. In those instances, the ER is referred to as "rough ER." In places where there are no ribosomes attached, the endoplasmic reticulum is referred to as "smooth ER." Smooth ER acts like an intercellular highway, moving molecules from one end of the cell to the other. The ER is primarily the site of protein synthesis.

Another type of organelle is called the Golgi body, or Golgi apparatus. The Golgi body looks like a series of flattened sacs. It's the processing and packaging plant of the cell. The Golgi body processes the cell's proteins and lipids before sending them out to other areas.

Yet another organelle is the lysosome. The lysosome looks like a sac, and it's full of enzymes that are used by the cell for digestion. The lysosomes break down particles of food and make them available for use by the cell. Depending upon the organism, the lysosome could be breaking down a banana, some grass, or whatever the organism uses for nourishment. Lysosomes aren't just confined to the digestive tract. Each and every cell needs nourishment. Once nourishment gets delivered to a cell (and, in the case of humans, your bloodstream delivers nutrients out to every single cell in your body), the lysosomes go to work breaking down the nourishment into a form the cell can use.

Now we come to the real powerhouses of the cell, the mitochondria. The mitochondria are where energy is stored and released. The released energy is used to form adenosine triphosphate. Adenosine triphosphate, or ATP, is just a storage vehicle for energy until it's used. It provides the chemical energy needed to drive the chemical reactions of the cell.

A typical animal eukaryotic cell contains many organelles within its cytoplasm.

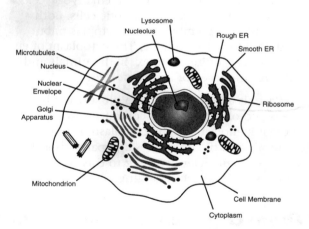

Mitochondria are not just your typical organelles. They have their own DNA, which allow them to divide and produce new mitochondria. In fact, scientists think that mitochondria developed from prokaryotic cells that moved into eukaryotic cells. According to the theory, the prokaryotic cells (that is, the fledgling mitochondria) sought to protect themselves from outside forces. In return for the protection they received, they began supplying energy to the eukaryotes.

The nucleus is where the cell's nucleic acids are synthesized. It directs all the activities of the cell. Within the cytoplasm, the nucleus is surrounded by a double membrane, called the nuclear envelope. Inside the nuclear envelope, the nucleus has its own version of cytoplasm: a dense protein-rich, jelly-like substance called nucleoplasm. The nucleoplasm contains fine strands of chromatin, a combination of DNA and proteins, and the nucleus.

Other things can be found in a cell's cytoplasm, but these organelles are the main "inhabitants."

The Basics of Plant Eukaryotes

Plants are also composed of eukaryotic cells. Nearly all the parts of the animal cell described previously can be found in plants. But plants also possess some additional structures.

In plants, a *cell wall* surrounds the plasma membrane and helps support and protect the plant. Plants also have *vacuoles* and *plastids*.

A plant's cell wall is made up of long chains of cellulose (a tough fiber that humans use to make paper and textiles; essentially, wood is cellulose), embedded with hardening compounds such as pectin (a carbohydrate that is usually found in fruits). Plants produce two types of cell walls: primary cell walls are formed during cell growth, and secondary cell walls are formed after growth has ceased. Both types strengthen the cell. This strengthening through cellulose and pectin (neither of which is found in the cells of eukaryotic animals) is a distinguishing feature of plant cells.

In plants, vacuoles are the equivalent of lysosomes in animal cells. The vacuoles store enzymes and waste products. In a mature plant cell, a vacuole can occupy up to 90 percent of the volume, often crowding other structures into the edge of the cell.

Plants use the sun as a source of energy. In a process called photosynthesis, plastids convert solar energy into chemical energy that the plant can use. Photosynthesis is a series of chemical reactions that convert radiant energy from the sun into chemical energy that the plant can use. The energy is then stored in the bonds of organic compounds, usually carbohydrates. Plastids really set plant eukaryotic cells apart; if eukaryotic animal cells had plastids, we wouldn't have to eat—we'd just engage in photosynthesis.

Bio Buzz

A **cell wall** is a rigid covering of a plant cell. A **vacuole** is an organelle that stores enzymes and waste products. A **plastid** is an organelle that converts solar energy into chemical energy.

Weird Science

Some of the waste materials vacuoles store is toxic to the rest of the cell. The vacuoles of some acacia trees store cyanide as a poisonous defense against plant-eating animals.

Three types of plastids work together in the photosynthesis process: Chloroplasts contain chlorophyll, a green pigment that absorbs sunlight. Chromoplasts synthesize and store pigments, such as orange carotenes, red pigments, and yellow xanthophylls. Like chloroplasts, some of these chromoplasts also trap sunlight for energy. Leucoplasts store food, including lipids, proteins, and starches.

Fluid Movement: The Key to Cell Interaction

In both plant and animal cells, the plasma membrane gives the cell its shape and flexibility and separates the internal structures from the environment. However, cells *do* interact with the surrounding environment. (How else would they obtain nourishment or make proteins, otherwise known as organic compounds?)

Bio Buzz

Because the plasma membrane selectively bars molecules from entering or leaving the cell, scientists refer to it as a **selectively permeable membrane.**

Cells are fluid; they're not hard. Even plant cells that are surrounded by hardening cellulose, or prokaryotic cells surrounded by a semirigid wall, still have things moving in and out of the cell, thanks to the plasma membrane. The plasma membrane is a complex barrier. It acts like a strict gatekeeper, or bouncer: Because it is fluid and moves, it allows some molecules into the cell and keeps others out, which is why scientists refer to it as a *selectively permeable membrane.*

The fluidity and selective permeability of the plasma membrane are what makes the cell work. The right molecules come in, and the right molecules go out.

Movement through the membrane can be accomplished in several ways: Diffusion, osmosis, and active transport are just a few of the ways molecules move in and out of the cell.

Diffusion is the random movement of molecules. The molecules move from an area of high concentration into an area of lower concentration. Placing a drop of ink in a glass of water is the best illustration of diffusion: The ink molecules start out concentrated and dark, and in a few seconds they're pale and sparsely concentrated.

Osmosis is the movement of water from an area of high concentration to an area of low concentration. Osmosis is actually a type of diffusion, but osmosis involves only water molecules and nothing else.

Diffusion and osmosis don't require the cell to expend any energy. But if a molecule moves through active transport, the cell has to expend energy. In *active transport*, a protein will move a certain material right across the membrane from an area of lower concentration into an area of higher concentration. It's a bit like a piggy-back ride. Because the protein is going against the flow, so to speak, the cell has to expend energy.

Cell Division

Most cells have a finite size. A large animal is not composed of one gigantic cell that just kept growing; it's composed of billions of cells that replicated, divided, and grew. With very few exceptions (your red blood cells being one of those exceptions), all the cells of living things go through a cell cycle. The cell cycle starts in the nucleus and is generally divided into two phases: interphase and mitosis.

Interphase is the first stage of the cell cycle. Here, the cell performs those functions that make it unique. (For instance, your eye cells are unique in relation to the cells in your teeth. When your eye cells go through interphase, they're ensuring that they'll remain eye cells—unique eye cells.) Essentially, it organizes and replicates all its DNA material so that there will be an exact copy for the new cell.

Interphase starts when the fine strands of chromatin (floating in the nucleoplasm of the nucleus) curl up and condense. This process of condensing forms a chromosome. Chromosomes are formed almost entirely of DNA and are vital because they contain all the genetic information of the cell.

Interphase is actually a three-part process during which each chromosome replicates itself. Let's look at humans as an example. Humans cells have 46 chromosomes. The first step (when the chromatin condenses into a chromosome) creates these 46 chromosomes. During the second step, the 46 chromosomes become 92. Then, during the third step, all 92 chromosomes organize themselves into lines and get ready to divide themselves in half. Replicating and then dividing in half is the whole point; it's how new cells are formed and how these new cells are exactly like the old ones. It's how new eye cells become eye cells and not tooth cells.

Think About It!

To learn more about movement in cells, read Mary Murray's article in *Discover*, March 1991, "Life on the Move." It's a great piece that describes how scientists observe cells.

Bio Buzz

There are many ways for molecules to move across the plasma membrane. **Diffusion** is the random movement of molecules across the membrane, from a high to a low concentration. **Osmosis** is just like diffusion, but it involves only water molecules. In **active transport,** protein goes against the flow and deliberately moves specific things through the membrane.

Dividing in half is the next stage of the cell cycle: mitosis. During mitosis, proteins organize themselves into a series of fibers, called a spindle. Spindle fibers are specifically constructed for each mitosis and are disassembled (by the cell) at the conclusion of each mitosis (the conclusion of that particular cell cycle). The proteins construct the spindle using microtubules, a material found in the cytoplasm.

During mitosis, the chromosomes form visible threads (well, visible if you have a good microscope). They organize themselves in the center of the cell (all 92 of them, in humans), and then they separate and (in humans) 46 chromosomes move into each newly formed cell.

To actually produce a new cell, the plasma membrane pinches into the cytoplasm and divides it into two parts, a bit like taking a balloon and twisting it into two sections; the volume and containment are roughly the same, but now there are two balloons instead of one. Scientists refer to this action as cell cleavage. During cell cleavage, the original cell pinches and divides, forming two new daughter cells, which represent the end product of mitosis.

Cell replication and division is complex and fascinating. Even though scientists have been able to observe the interiors of cells and to figure out just how the cell cycle works, they still don't know the answer to the *why* of life, which still remains a mystery.

Weird Science

The word *mitosis* comes from the Latin stem *mito*, which means "threads." When scientists first observed mitosis, more than a century ago, they described "threads" within the cells. So, mitosis was the name they came up with to describe the movement of those threads.

But Can You See It?

As cell theory developed, it became obvious that verbal or written descriptions of size were woefully inadequate. Van Leeuwenhoek's "a thousand times smaller than the eye of a big louse," just didn't cut it (scientifically speaking) when it came to describing cells. Viewing things through the lens of a microscope required a whole new approach to relative size. Viewed through the lens, an organism could appear large enough to hold in the palm of your hand. In reality, the organism was invisible to the naked eye.

Try It Yourself

You can see how cells—specifically, chloroplasts—move about with only a simple aquarium plant, a bright light, and a student microscope.

Place a sprig of *Elodea*, a simple aquatic plant found in most aquarium stores, in a tank or a bowl of water. Shine a bright light on it for at least an hour. Then examine a leaf of the plant under a microscope. Are the chloroplast cells moving in one direction? Are they moving at all? Do you think that the movement of these organelles helps the cell to function?

By the mid-1800s, science had become an international enterprise. All over the world, scientists were working on projects and comparing notes with each other. It was obvious that an internationally accepted system of measurement was needed.

Enter the newly developed metric system. The Paris Academy of Sciences devised the metric system in 1791, and the French National Assembly adopted it in the same year. The beauty of the system was that it was extremely easy to use.

Every unit is based on a multiple of 10, which makes it easy to change from a small unit to a large one. Today, the system is called the SI system (after the French term, *Systeme International*) and is regulated by an international commission.

In scientific work, the basic unit of measurement is the meter. For reference purposes, a meter is about 10 percent longer than an English yard. The meter is divided into 10 decimeters, which are in turn divided into 10 centimeters, and a centimeter is divided into 10 millimeters. A millimeter is one-thousandth of a meter. It keeps declining from there.

You can see where a steadily declining system of measurement would be useful when it comes to microscopic organisms. You could measure anything. For example, a human egg cell is approximately one-tenth of a millimeter in diameter. Because a human egg cell is quite large by microscope standards, the system goes even smaller. A micrometer is one-thousandth of a millimeter, and a nanometer is one-billionth of a meter.

The metric system did away with descriptions like "a thousand times smaller than the eye of a big louse" and replaced them with precise statements that could be interpreted by anyone.

We've covered a lot of ground in this chapter, as well as a lot of complex material. Cells are tiny, but they are indeed the foundation of life. The first cells on Earth, that first spark of life, were simple prokaryotic cells. We discussed the basics of more complex cells, known as eukaryotic, and examined how these cells replicate and reproduce, and how scientists are able to see cells and cellular functions. In the next chapter, we'll be talking about how these cells (and the larger organisms they make up) actually function.

The Least You Need to Know

➤ Cells are the smallest unit of life. They are the building blocks of all living organisms.

➤ All cells have a plasma membrane, cytoplasm, and a nucleus or nucleoid.

➤ Living organisms develop through cell division and growth.

➤ Scientists observe and measure cellular life forms using complex microscopes and the SI system (the metric system).

Physi-What?

How do two single cells turn into Michael Jordan? How do two single cells turn into a platypus? How does a redwood get to be 100 feet tall? The answer is energy. Energy is the real key to development and growth of a cell.

To accomplish internal functions or to interact with its environment, a cell has to have energy. It must capture energy, convert it, and use it, or nothing happens (well, death happens, but that's not the topic of this book). In this chapter, we'll explore the energy of life.

Physiology: Movin' Metabolism

Physiology is the branch of life science that studies the way plants and animals carry out the various functions necessary to life. Physiology deals with how an organism works and interacts with its environment. A physiologist might ask the questions such as "How does an amoeba move?" or "How does a muscle lift weight?"

Bio Buzz

Physiology is the study of living organisms and the ways they carry out the functions necessary for life.

Given the number of living organisms on the planet, the scope of physiology is huge. And if the number of organisms wasn't enough, their functions and interactions certainly are. Because physiology involves not only the study of organisms as a whole, but also the actions of individual organ systems and the functions of cells and parts of cells, the field of study is virtually limitless.

To get an overview of the world of physiology, let's take a look at meerkats, those cute, furry African mammals. (Think Timon in *The Lion King*.)

After a freezing night on the Kalahari Desert, these little animals emerge from their burrows, stand tall, and face the rising sun. Without realizing it, they're helping their internal organs to function better. If their internal temperature drops too low, which it would if they didn't warm up after spending a freezing night underground, their rate of enzyme activity would drop sharply and their metabolism would suffer.

After warming up, the meerkats start to forage for the day's food. Unfortunately, the farther they get from their burrow, the more vulnerable they are to predators, but their hunger outweighs their caution. If they managed to eat only very little the day before—say, a bunch of tiny insects—they need more glucose, the nutrient required by all their cells, including brain cells, in order to function.

On the other hand, if they ate a chubby lizard yesterday, they may have been able to take in more nutrients than their cells needed. In that case, their bodies converted the excess energy into fat and stored it. After settling in for the night, their bodies automatically triggered a shift in the type of molecules used to support cell activity while they slept. The stored fats were broken down, transported by the bloodstream to the liver, and then broken down even further into glucose, which was then delivered to all the cells. In either case, the energy the meerkats took in allowed a host of physiological activities to take place. Their circulatory systems picked up the converted nutrients and transported them to all the cells. Their respiratory systems helped cells use the nutrients by supplying them with oxygen for aerobic respiration. As the meerkats wake and face another day of foraging, thirst, and heart-pounding flight from predators, their nervous and endocrine systems are working together to get them through the challenges of existence.

Although we're talking about meerkats now, cell functions and interaction really applies to all living organisms, from meerkats to humans, to floating diatoms in the ocean (single-celled photosynthetic organisms that produce most of the oxygen used on our planet).

The term *physiology*, like *biology*, encompasses a great many different levels of study. When it comes to human physiology, for instance, you can discuss neurophysiology (the study of the functioning of the nervous system), cardiovascular physiology (the study of the functioning of the heart and blood vessels), and a host of other physiological systems.

Try It Yourself

Over the centuries, science has advanced because people thought and pondered things of interest. Many questions are not immediately answerable, and some may never be. Put on your thinking cap and see if you can advance the scientific cause.

When dust, pollution, smoke from a forest fire, or volcanic activity block sunlight, there are some very direct effects on the entire energy cycle. What short-term, long-term, and far-reaching effects can you think of? What is affected?

Most scientists are convinced that the Earth's early atmosphere contained very little oxygen, yet life developed and progressed. What type of metabolism might these early organisms have had? How did respiration work billions of years ago?

Then there's comparative physiology, which studies the similarities and differences between the methods different organisms use to solve the problems of existence. General physiology seeks to integrate the findings of all the other branches of physiology into one coherent picture. (If you're looking for a new career, physiology may be it. You'll never run out of material!)

Obviously, physiology is a huge field. But all physiological research includes a central theme, even if it's indirect: the use of energy. No system can function and no organism can interact without energy. Where does the energy came from that sets all these in operation?

Weird Science

Lest you think the topic of physiology is a yawner, consider the case of the murdering mushroom. Oyster mushrooms actually attack unsuspecting nematodes (worms) with a toxin and then send out filaments to grow right into the stunned worm, actually digesting it from the inside out. And, yes folks, physiologists research this stuff—and get paid for it.

Photosynthesis

All living cells need energy to function. They use energy to digest food, grow, build molecules,and reproduce. The original source of all this energy is light. Without sunlight, everything would shut down. Think about it: You personally may not need or use sunlight, but you have to eat to stay alive. Even if you're a pure carnivore—vegetables never pass your

Think About It!

Are plants the only organisms that engage in photosynthesis? For an interesting perspective on the topic, read B. Davis's article, "Going for the Green," on page 20 of the February, 1992 issue of *Discover*. It describes work on artificial systems for photosynthesis.

Bio Buzz

The release of chemical energy for cellular functions is called **respiration.**

lips—the meat you're eating got to your table only because it ate plants. And without sunlight, the plants wouldn't be here fueling the entire planet with energy (directly, in the case of your meat, and indirectly, in the case of you consuming the meat).

For a cell to obtain usable energy (especially your cells), the radiant energy of light must be converted through a series of complex chemical reactions. The first stage of this process is called photosynthesis, as we discussed in the previous chapter.

As discussed, carbohydrates such as sugars and starches are the primary source of energy for most living organisms. The energy is stored in the bonds that hold the molecules of a carbohydrate together. The release of this chemically stored energy, known as *respiration,* is the second step of the energy process. Most organisms rely on the processes of photosynthesis and respiration to provide all their energy needs, either directly or indirectly.

In life science, there are lots of ways to classify organisms. But one of the primary classifications is based on the origin of the organism's food energy. Plants and a few other organisms can make their own food, and they do this by turning inorganic material (such as light) into organic material—specifically, glucose. Such organisms are known as called *autotrophs*.

All plants are autotrophs because they convert the sun's radiant energy into chemical energy through the process of photosynthesis. So, for the purposes of energy classification, a 10-story California redwood is no different than a single-celled diatom drifting in the Atlantic.

Animals and other organisms can't make their own food like plants can. Known as *heterotrophs,* they get their energy by feeding on autotrophs or by eating other heterotrophs that have fed on autotrophs. Alfalfa is an autotroph. A cow is a heterotroph that feeds directly on autotrophs. Humans are heterotrophs that feed on other heterotrophs (the cow) in addition to feeding on autotrophs (lettuce and broccoli). Ultimately, all life on Earth depends on autotrophs for energy.

Photosynthesis and respiration are *biochemical pathways*. Together, they form an energy-producing cycle. Using radiant light, autotrophs produce carbohydrates and oxygen (the process of photosynthesis). Heterotrophs and other autotrophs use the carbohydrates and oxygen, in turn producing carbon dioxide and water (the process of respiration), which the original autotrophs then use to produce more carbohydrates and oxygen.

In photosynthesis and respiration, the energy is frequently transferred from one reaction to the next in the form of the energy storage molecule called adenosine triphosphate, or ATP. In order to release the energy from storage, the bonds that form ATP must break and reform into another molecule called adenine diphosphate, or ADP. A The breaking of the bond produces the chemical energy. Although cells must use energy to break the bonds of ATP, more energy is released than is used up, so overall, the formation of ADP from ATP is an energy-producing reaction.

Photosynthesis, respiration, and the ATP-ADP cycle form a complete energy-producing sequence. The process of photosynthesis stores energy in organic compounds, respiration releases this energy, and the freed energy forms ATP. The breakdown of ATP into ADP produces the energy that fuels cellular activity.

Photosynthesis, the start of the cycle, can be divided into three sets of reactions:

1. Light absorption by pigments, such as chlorophyll
2. Light-dependent reactions
3. Light-independent reactions

Let's take a look.

Chlorophyll: The Start of It All

Chlorophyll are actually light-absorbing compounds, or pigments, that are found inside chloro-plasts. Chloro-plasts are organelles found within the cytoplasm of plant cells.

Chloroplasts are complex organelles, which is not surprising, because the fate of the world rests upon them. Each chloroplast has three membrane systems, inner, middle, and outer. When it comes to photosynthesis, the inner membrane is what we're interested in.

Bio Buzz

An **autotroph** is an organism that uses energy, usually from light, to produce organic molecules from inorganic substances.

Bio Buzz

A **heterotroph** is an organism that cannot produce its own energy. Heterotrophs obtain energy by eating autotrophs or other heterotrophs that have eaten autotrophs.

Bio Buzz

A **biochemical pathway** is a series of complex chemical reactions. Biochemical pathways produce chemical products that are reused. Both photosynthesis and respiration are biochemical pathways.

59

Weird Science

When it comes to planetary health, the importance of autotrophs—those organisms (primarily plants) that create their own food from inorganic material—cannot be overstated. Autotrophs are the basis of the biochemical pathway that keeps all of us alive. When scientists fret about the demise of the rainforest or the fate of oceanic diatoms, they're not just whistling *Dixie*. They're talking about the foundation of one of the most critical biochemical pathways on Earth. Without autotrophs—healthy ones—you wouldn't exist.

Weird Science

If you lined up 2,000 chloroplasts, they would stretch no longer than the diameter of a dime. How many chloroplasts do you think are in a single leaf of lettuce? That gives you an idea of the magnitude of all the metabolic events required to feed this planet, doesn't it?

The inner membrane is similar to a cell's plasma membrane. It's a double layer of lipids embedded with proteins. This membrane is more complex, though, because it's arranged in thick "stacks" called grana. The grana are surrounded by a protein-rich solution called the stroma. And each granum is composed of flattened, membranous sacs called thylakoids. The chlorophyll molecules are located on the thylakoids. It's a bit like "the house that Jack built." If your eyes are starting to glaze over, remember that we're talking about only *one* membrane here, the inner membrane of a chloroplast molecule. Fortunately for you, this one membrane is where all the action takes place: light absorption, light-dependent reactions, and light-independent reactions.

Thylakoids contain several types of light-absorbing pigments, of which chlorophyll are the most abundant. Chlorophyll mostly absorb red and blue light, but all chlorophyll reflect green light, so organisms containing chlorophyll look green. Other pigments trap yellow and green light and transfer it to the chlorophyll. The absorption of light by chlorophyll is the first step of photosynthesis.

Light-Dependent Reactions

The next step of photosynthesis, the light-dependent reaction, converts the light energy into chemical energy. A light-dependent reaction actually occurs in two stages. In the first stage, the solar energy is converted into electrical energy, which consists of a flow of electrons. In the second stage, the electrical energy is converted into chemical energy that is stored in chemical bonds.

When solar energy strikes a chlorophyll molecule, the electrons within the molecule become excited. As the excitement level rises, the electrons start moving from the high-energy area into lower-energy areas. This methodical movement of electrons is called an electron transport chain.

The loss of electrons creates unstable chlorophyll molecules. Through a series of complex interactions, molecules break down to form compounds, water among them. To regain stability, the chlorophyll molecules take electrons from the water molecules. But this breaks up the water molecule, releasing hydrogen ions and oxygen gas. The loose hydrogen ions form into different compounds, with ATP being one of them.

Through this two-step process of light-dependent reactions, solar energy is converted into chemical energy. The chemical energy is held in the bonds of the ATP and other compounds.

Light-Independent Reactions

The next step of photosynthesis is to take the chemically stored energy and use it to form organic compounds. During light-independent reactions, inorganic one-carbon molecules are bonded, or "fixed," into organic three- and five-carbon molecules. These reactions are carried out through the *Calvin cycle*.

The Calvin cycle goes in (you guessed it) cycles. One full turn of the cycle fixes one CO_2 molecule and regenerates an RuBP molecule. Three full turns produces one PGAL molecule. (Don't worry about RuBP or PGAL molecules; we're only talking about the action of the cycle here, the process.) Six full turns produces a six-carbon sugar, such as glucose, the real source of energy for most living organisms.

There are other ways plants fix carbon, but the Calvin cycle is the most common method of fixing carbon and forming carbohydrates.

Weird Science

The **Calvin cycle** is named after Melvin Calvin (1911–present), a botanist at the University of California at Berkeley who figured out the process. Most plant species in temperate climates use the four-step Calvin cycle to fix inorganic carbon into organic molecules.

Respiration: Breathing?

So what happens to all this energy? How do cells actually use all this "converted" energy? Let's take a look.

Take a deep breath. You've just filled you lungs with oxygen-rich air. Now exhale. You've just gotten rid of waste carbon dioxide. You've also just experienced external respiration.

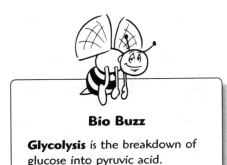

Bio Buzz

Glycolysis is the breakdown of glucose into pyruvic acid.

External respiration is something with which all humans are familiar. Along with food, air provides us with energy. We inhale "energy" (primarily in the form of oxygen) and exhale waste products (primarily in the form of carbon dioxide). Cellular respiration is different from external respiration in many respects, but it also involves energy and waste products.

Cellular respiration is a series of chemical reactions that occur inside the cells of every living organism. During cellular respiration, cells break down food molecules. When food molecules break down, they release the energy held in their bonds for the cells to use.

Cellular respiration can be broken down into two major categories: anaerobic respiration and aerobic respiration. Each type of respiration is a different biochemical pathway. Both are different, yet both start with the same process: glycolysis, or the breakdown of sugar.

Break Down That Sugar!

Glycolysis occurs in the cytoplasm of cells and can occur with or without help from oxygen. The main result of glycolysis is that one six-carbon molecule of glucose gets split into two three-carbon molecules of pyruvic acid, which is a form of energy that the cells can use.

The process of glycolysis produces energy and sets up the cell for the next phase of respiration, either anaerobic or aerobic respiration.

Anaerobic Respiration

Anaerobic respiration is the type of cellular respiration that occurs without oxygen and ultimately produces either lactic acid or alcohol. Lactic acid doesn't generate any of those useful ATP molecules, but it does generate another molecule (NAD+), which is used to produce ATP.

Ever wonder why your muscles hurt after exercising? When your muscles work hard, they run out of oxygen and have to use lactic acid to get more energy. It's the lactic acid that makes your muscles sore.

Alcoholic fermentation is a process some plants and unicellular organisms use to convert pyruvic acid (the usable breakdown of glucose) into ethyl alcohol. The beer and wine industry loves those little ethyl alcohol producing organisms (also known as yeasts). The main reason that alcoholic beverages are so high in calories is because the energy from the glucose and pyruvic acid molecules isn't used to form ATP molecules; it remains in the ethyl alcohol.

Aerobic Respiration

Glycolysis doesn't require oxygen, but aerobic respiration does. Aerobic respiration is glycolysis, followed by the breakdown of pyruvic acid, which requires oxygen. Aerobic respiration involves three steps, and all of them occur in the mitochondria. (Remember those complex little organelles?) But the most important step is the second one, called the Krebs cycle, which is a biochemical pathway that produces ATP molecules. The Krebs cycle is named after Hans Krebs, a German-English scientist who discovered the biochemical pathway.

Through a series of complex chemical interactions, the Krebs cycle, in combination with glycolysis, ends up producing 4 ATP molecules and 12 carrier molecules. The carrier molecules are used to carry electrons from the Krebs cycle to the electron transport chain (sounds suspiciously like photosynthesis, doesn't it?), where water and more ATP molecules are formed.

The electrons travel from carrier to carrier, moving down through lower energy levels until they finally come to oxygen. The oxygen then combines with hydrogen to form water. But without oxygen, this final step of aerobic respiration, the electron transport chain, could not occur.

Both types of respiration, anaerobic and aerobic, free up energy for cells to use. Scientists have figured out that each glucose molecule is capable of producing 38 ATP molecules through respiration. Because anaerobic respiration produces no ATP molecules (only the ATP molecules generated by glycolysis are available for cells engaged in anaerobic respiration), aerobic respiration is about 19 times more effective when it comes to providing energy for cellular activities.

More Building Blocks for Energy

DNA and RNA do more than just control the genetics and replication of cells. DNA controls the production of proteins within the cell. In turn, these proteins form the structural units of the cell and control all the chemical reactions within the cell. Proteins are the true building blocks of cells, and they wouldn't exist without DNA and RNA. (Without cells, life as we know it wouldn't exist.)

DNA: The First Building Block

DNA does much more than just control the passage of genetic material to new cells; it's also responsible for the production of protein, a basic component of all cells. DNA doesn't create proteins directly. DNA

Think About It!

DNA is complex and fascinating, and the story of its discovery is no less so. For a firsthand account, read James Watson's book, *The Double Helix* (W. W. Norton, 1981). Watson (along with his colleague, Francis Crick) discovered the structure of DNA.

creates RNA and lets the RNA get on with making the necessary proteins. But because RNA is produced by DNA, you still get back to the basic importance of DNA within any cell. DNA stores the information needed for protein synthesis, and RNA transmits and uses this information to produce proteins.

RNA: The Next Building Block

The primary purpose of RNA is to synthesize, or create, proteins. Like DNA, RNA is composed of nucleotides, but instead of two strands intertwined in a double helix, RNA has only a single strand of nucleotides.

RNA exists in three different structural forms, and each form plays an important role in the synthesis of protein. Messenger RNA (mRNA) is a single, uncoiled strand used to transmit the basic information from DNA. Transfer RNA (tRNA) is a single strand of RNA folded back upon itself. This allows complementary bases to pair up and bond with a specific amino acid. Ribosomal RNA (rRNA) is RNA in globular, not strand, form. Within any cell, rRNA is the major constituent of ribosome organelles. The actual function rRNA performs during protein synthesis is not really understood. (From a future career standpoint, this is fortunate—you can be the scientist to figure it out, reaping praise and fame in the process.)

Bio Buzz

Protein synthesis is the formation of proteins by RNA, informed by information coded by DNA.

Bio Buzz

A **codon** is a specific group of three sequential bases of messenger RNA (mRNA). Codons recognize and attract specific amino acids, using transfer RNA (tRNA) as an intermediary.

Protein Synthesis

Plants use photosynthesis, but animals use *protein synthesis* to create the energy necessary to sustain life. Depending on its complexity, an organism will have from several hundred to several thousand different types of proteins in its makeup.

Proteins are composed of one or more polymers called polypeptides, and each polypeptide consists of a very specific sequence of amino acids. The amino acids are linked together by peptide bonds (and you thought protein was just chicken or steak).

There are 20 different kinds of amino acids, and a protein can consist of hundreds or thousands of amino acids. All these amino acids must be arranged in a particular sequence, or the protein won't function properly. All the structural and functional aspects of a protein are determined by its amino acid sequence.

The assembly of all these amino acids into the correct sequence is the task of protein synthesis. But before you can even get to the process of protein synthesis or the assembly of all these amino acids, we need to take another look at nucleic acids.

The genetic code contains all the information needed for proper cell functioning. Because DNA makes RNA, and because RNA makes the proteins, the information needed to assemble the amino acids in the proper sequence is coming from DNA.

Remember mRNA? Using DNA as a template, the mRNA is transcribed with all the correct information. The whole genetic code of DNA is reflected in the sequence of bases in mRNA. A specific group of three sequential bases of mRNA is called a *codon*. A codon recognizes, or codes, a specific amino acid, using tRNA as an intermediary.

There are 64 possible codons. Not all codons deal with amino acids. Some codons are start signals that tell a ribosome to start reading an mRNA molecule. Other codons tell the ribosome to stop reading the mRNA molecule.

Physiologists code all amino acids in the same way for every organism. (We're back to that "all life comes from an original cell" theory.) For instance, UUU (the coding for the amino acid phenylalanine) is the same in humans, mice, or bacteria. Likewise, AUG is the universal start code for every organism. Virtually every organism uses the same genetic code. (The next time you refer to your ex-boyfriend as a piece of bacteria, you won't be too far off the mark—they've got a lot in common.)

The process of actually assembling protein molecules (using the information encoded in the mRNA) is called translation. Translation begins when mRNA moves out of the cell's nucleus and migrates to a group of ribosomes. The ribosomes are where the actual synthesis of protein takes place.

Up until now, amino acids have been floating freely in the cytoplasm. Transfer RNA (tRNA) transports these free-floating amino acids to the ribosomes, where they pair up with the correct codons on the mRNA. When the tRNA reaches the ribosomes, assembly starts.

Protein synthesis occurs throughout the cell. Several different ribosomes can be simultaneously translating the same mRNA. As a ribosome moves along a strand of mRNA, each codon is correctly paired with its anticodon, or matching counterpart picked up by the tRNA, and is added to the peptide chain. An enzyme in the ribosome catalyzes a chemical reaction that binds each new amino acid to the developing chain. The process continues, with each new amino acid being assembled and bonded to the chain until the ribosome eventually reaches the "stop" codon. This ends the whole process, the mRNA is released, and the polypeptide chain is complete.

And, voila! Protein has been created! The creation of protein is important because that's what cells are essentially made of. Your hair, for instance? It's basically protein (at least in its cellular makeup). Some cells (like muscle cells) have more protein than others, but all of them have it.

Without the ability to capture, release, and use energy, cells wouldn't last long after their creation. They'd be a one-second wonder that instantly perished. Through photosynthesis, plant cells are able to sustain themselves (and, indirectly, everything else on the planet). Through protein synthesis, other organisms are able to sustain themselves. Creation is just dandy, but sustenance ensures the furtherance of that creation. It all comes down to energy: Use it, or lose it—the "it" being life.

We've covered a lot of complex, abstract material in this chapter (just think about all the exercise your brain has gotten). We've talked about how cells procure and produce the energy needed for all their activities and functions. In the next chapter, we'll be talking about all the diverse life forms on this planet, how scientists classify them, and how the most simple group—bacteria and viruses—functions.

The Least You Need to Know

➤ Physiology is the study of living organisms and the way they carry out the functions necessary for life.

➤ The original source of the energy cells use is light. The radiant energy of light must be converted into energy that the cell can use.

➤ Photosynthesis is the first step in converting radiant light energy into usable energy. Photosynthesis is the process of converting radiant light energy into chemical energy that is stored in the bonds of organic compounds, usually in carbohydrates.

➤ Proteins are the building blocks of all cells. Proteins are constructed through DNA and RNA.

Part 2

The Kingdoms of Life

Our planet is a lot of things. Although it has a lot in common with the other planets in our solar system—its shape, its composition—what really sets our planet apart is life. And earth doesn't just have life on it, it's got LIFE on it. Millions and millions of diverse life forms.

How do scientists even begin to classify and study all these different forms? How do you distinguish a butterfly from a moth or a bird—let alone another butterfly? Without some general groupings, scientists would never get anywhere.

Scientists use five general groupings referred to as "kingdoms." Every known organism can be classed into a kingdom—from one-celled bacteria, to yourself. Each kingdom appears vastly different from the next. (Why wouldn't it? How could a rhino be like a mosquito?) Still ... anything that's alive has some unifying traits. Life is nothing if not organized. Just take a look at the next few chapters and you'll see what we mean.

Chapter 6

Watch Out! Viruses and Bacteria

In This Chapter

➤ Five kingdoms, five billion organisms

➤ Organisms without a kingdom: viruses

➤ Viral structure and diversity

➤ Bigger than a virus, smaller than a breadbox: bacteria

➤ Bacterial types and structure

Early scientists, like Aristotle, divided living things into two classifications: plant or animal. But plenty of organisms don't fit in either of those groups. What about the virus that has you flat on your back for a week? It's definitely a living organism, but is it a plant or an animal?

Modern scientists have created several categories (aside from plant and animal) to classify living organisms. Some of these classifications, such as viruses and bacteria, might not contain what you'd consider to be "real" life, but they are—every single one of them. That's what we'll discuss in this chapter.

In this chapter, we'll be talking about the five kingdoms, the starting point for classifying any organism. We'll also be discussing viruses, an organism that lives in no-man's land and doesn't really fit into any kingdom. In the rest of the chapter, we'll be talking about bacteria, the real inhabitants of the first kingdom: the Kingdom Monera.

The Five Kingdoms

Most modern scientists use a five-kingdom system for classifying living organisms. But keep in mind that the five-kingdom system is only the latest (and currently accepted) method of classification. Even as you read this, scientists are developing other methods of organizing the diversity of life.

The Kingdom Monera

The Kingdom *Monera* (sounds like a sci-fi planet, doesn't it?) contains the least number of species of any of the kingdoms (only 5,000 identified species) and the greatest number of living things on Earth. That's because nearly every moneran species has billions of individual members.

Some common, or at least recognizable, monerans are *Escherichia coli*, the bacteria that are normally present in your own intestines, and *Clostridium tetani*, the bacteria that causes tetanus.

Bio Buzz

Monera are prokaryotic organisms that lack a nucleus and other membrane-bound organelles.

Monerans are prokaryotic organisms, which means that their cells lack nuclei and other membrane-enclosed organelles. Monerans have the most basic of all cell structures. Many of them live in aquatic and terrestrial habitats and obtain their nourishment primarily through absorption. Others are autotrophs, producing their own food, usually through photosynthesis. Moneran species reproduce asexually (they don't need a partner of the opposite sex). They produce their own genetic material and then split in two (and then those two split in two, and so on).

The Kingdom Protista

Like the Kingdom Monera, a lot of the Kingdom Protista (which contains 50,000 species) is composed of bizarre organisms you probably wouldn't recognize. But there's one major category you'd definitely recognize. Seaweed, the "plant" that twists about your legs when you're swimming in the ocean, is not a plant at all; it's a protist.

Protists are more complex than bacteria. All of them are eukaryotic organisms (which, if you recall, means that they have a more complex cell structure). But they lack any specialized tissue systems. Protists have nuclei and membrane-enclosed organelles, but not a lot else. Protists can be either unicellular or multicellular.

Bio Buzz

Protists are eukaryotic organisms with nuclei and membrane-bound organelles, but they lack organized tissue systems.

Protists live in aquatic or very moist environments. They obtain all their food by absorption, ingestion, or photosynthesis. Protists can reproduce either sexually or asexually.

The Kingdom Fungi

Now we're getting into stuff you can definitely recognize: mushrooms, rusts, and bread molds, to name a few. Anything in the Kingdom *Fungi* is composed of heterotrophic unicellular and multicellular eukaryotic organisms, which is a real mouthful—this essentially means that fungi have a complex cell structure (they're eukaryotes, with nuclei and membrane-bound organelles). They can be single- or many-celled, *and* they don't manufacture their own food: They consume either autotrophs or other heterotrophs. Most fungi absorb their nutrients rather than ingesting them, and most species are also terrestrial, which means that they live on land, not in the water or air. More than 100,000 species call the Kingdom Fungi home, including yeasts, puffballs, mushrooms, rusts, and bread molds.

Bio Buzz

Fungi are heterotrophic unicellular and multicellular eukaryotic organisms.

The Kingdom Plantae

The Kingdom *Plantae* is a kingdom whose inhabitants you can definitely recognize—especially because it's solely composed of (yes, you guessed it) plants. All plants are autotrophic and multicellular. Most of them are terrestrial, but some are aquatic. Plants reproduce both sexually and asexually.

There are more than 350,000 species in the Kingdom Plantae. Plants include conifers, deciduous trees, mosses, ferns, and flowering plants.

Bio Buzz

Plantae consists of all multicellular and autotrophic plants.

The Kingdom Animalia

Now we're at the top of the pyramid, at least in terms of complexity and the living things with which you're most familiar. The Kingdom *Animalia* is composed of eukaryotic, heterotrophic, multicellular organisms that obtain nutrients by ingesting

Bio Buzz

Animalia includes eukaryotic, multicellular, heterotrophic organisms that obtain nutrients through ingestion.

food. Animalia species are both terrestrial and aquatic. Nearly all of them reproduce sexually, but a few species can reproduce asexually.

There are more than 1,000,000 species in the Kingdom Animalia. Animalia includes everything from spiders to elephants, to mice, to crocodiles, to you.

The kingdoms are only the starting point for classification of organisms. From there, the system progresses (seemingly to infinity), starting with phylum, then class, then order, then family, then genus, and, finally, specific epithet. For most of our purposes, though, we'll be dealing only with the broader categories in this book.

Viruses: Organisms Without a Kingdom

As you can tell, the five-kingdom system classifies organisms in order of their cellular complexity. But a few living things still don't fall into any category. Take viruses, for instance. Viruses are interesting (I know you don't find the common cold very interesting, but scientists do) because they straddle the two worlds of living and nonliving things. Viruses are alive, but not quite the same way that other organisms are "alive."

Viruses lack most of the essential features that other living organisms possess. Namely, they don't possess organelles, or DNA *and* RNA, which brings us to their most fascinating feature: reproduction. One virus cell becomes many by taking over the genetic machinery of the host cell. That means that when that cell divides, the new cell also contains the virus. The release of the new virulent cells is what causes the spread of many infectious diseases in both plants and animals.

Virus Structure

A virus is a biological particle (which is a pretty open-ended description, but there you have it) that's composed of genetic material and protein. It's got the basic elements of life, but that's about all it's got. A typical virus will have either DNA or RNA encased in a protein coat called a capsid. Because viruses need a host cell to reproduce, they're called *obligate intracellular parasites*.

Viruses are constructed of compounds usually associated with living organisms (that is, genetic material and protein), but they're not considered to be living organisms. They have no cell membrane, nucleus, cytoplasm, or organelles. They can't carry out cellular functions, and they can't reproduce through either mitosis or meiosis. But they can take over certain cells within an organism and cause disease, and that's a pretty nifty trick for something that's basically a nonentity.

If a virus causes disease, it's said to be virulent. But if it doesn't cause disease immediately, then it's said to be

Bio Buzz

An **obligate intracellular parasite** can reproduce only by invading a host cell and using that cell's enzymes and organelles to reproduce.

temperate. You'll note that the presumption here is that a virus is going to cause disease at *some* point, even if it's not immediate.

Because they're not cells, invasion and takeover is the only way that viruses can do anything. Outside the host cell, the virus is a lifeless particle with no control over its movements. A virus "moves" in water, in food, on the wind, or through the blood and other bodily secretions of its host.

By the 1800s, scientists knew about bacteria because they could see them under microscopes. But they were puzzled because the cause of some illnesses didn't show up under the microscope. Something else out there was smaller than bacteria and could transmit disease. Whatever it was, it wasn't reproducing or behaving quite like bacteria.

As frequently happens, scientific intuition was right. But it wasn't until 1935, when Wendell Stanley (1904–1971) isolated the virus that causes tobacco mosaic disease that scientists learned what the "thing" was. Because viruses are indeed much smaller than bacteria, the development of the electron microscope was what really allowed scientists to start studying viruses.

Think About It!

Want to know more about viruses? Check out B. Geunno's article, "Emerging Viruses," in the October, 1995 issue of *Scientific American* 273(4) on pages 56 to 62.

Among other things, the electron microscope revealed the remarkable shapes of these clever creatures. When it comes to structure, most viruses fall into one of two shape categories: polyhedral or helical. A polyhedron is shaped somewhat like a soccer ball, roughly spherical and usually with 20 triangular-shaped faces. Polio, warts, the common cold, hepatitis, and other diseases are all caused by polyhedral viruses.

Another common shape for a virus is helical. Helical viruses include smallpox, rabies, influenza A, B, C, and other plant diseases. Helical viruses are rod-shaped. They have a helical strand of nucleic acid that's surrounded by a helically arranged protein coat.

Vaccinations, Pasteur, and Rabies

Louis Pasteur spent much of his life working with microorganisms. After winning the prize for disproving spontaneous generation, in 1864, he went on to research bacterial diseases in silk worms (saving the entire silk industry in the process), chicken cholera, anthrax (a cattle disease), and the processes of fermentation used in the brewing industry. He intended to devote the rest of his life to improving the French brewing industry, but he found himself sidetracked in the 1880s by the intriguing problem of rabies.

Based on his work with chicken cholera and anthrax, both of which had responded to weakened vaccines, Pasteur felt that perhaps a vaccine could also control rabies. At that time, vaccines were not commonplace. A vaccine is a preparation, usually composed of weakened bacterial or viral cells, that is used to induce the body to build up its own resistance to a specific pathogen, or disease-causing organism. Vaccines are usually administered by an injection, called a vaccination. Vaccinations are also referred to as inoculations.

In 1881, nearly 60 years old and semiparalyzed from a stroke, Pasteur began work on a vaccine for rabies, a deadly virus that infected dogs and other animals, but that was also fatal to most humans. He created the virus he would work with by first inoculating a healthy dog with the virus in order to grow a culture. Then he inoculated other dogs with the cultured virus. All the dogs he inoculated with the virus developed the disease. Those that recovered became resistant to subsequent exposures. This led Pasteur to hypothesize that a single infection produced immunity to the disease.

Through a series of shots, Pasteur gradually exposed experimental animals to the virus. Starting with dead viruses, each shot was more virulent than the last. Through this method, he was able to produce immunity in the animals without killing them.

Pasteur's rabies vaccine worked well on animals, but he was reluctant to try it on humans. Then in July 1885, physicians came to Pasteur, begging him to try the vaccine on a 9-year-old boy who would otherwise die. Pasteur administered the treatment, and the boy survived.

At the time of Pasteur's death in 1895, more than 20,000 people had been treated with "the Pasteur Treatment" for rabies. The death rate was less than one-half of 1 percent. Nowadays, Pasteur's rabies vaccine is a standard inoculation.

Complexity Defined: The AIDS Virus

During Pasteur's time, rabies was a real problem because it was a complex virus. Nowadays, other viruses are so complex that they fall into their own class. The virus that causes acquired immune deficiency syndrome, or AIDS, is such a virus.

The AIDS virus, more accurately known as the human immunodeficiency virus (or HIV), has two single strands of RNA in its core. Two layers of protein surround the RNA strands. A layer of lipids surrounds the inner protein layer. Glycoprotein molecules and proteins with sugar chains attached are embedded in the lipid layer. The lipid layer (with everything embedded in it) is what forms the virus's capsid. For sheer complexity, comparing the AIDS virus to the rabies virus is like comparing a Ferrari to a child's go-cart.

Someone who dies from AIDS isn't actually dying from a specific virus, per se. The AIDS virus replicates rapidly and attacks the T-cells in your body. T-cells are cells in your immune system that protect you from disease. AIDS essentially kills them, which leaves you vulnerable to a whole constellation of diseases that you normally wouldn't be susceptible to. It's those diseases (which have been able to invade, courtesy of the AIDS virus) that kill you.

Because viruses aren't really living things, they don't fall into any of the five kingdom classifications. Instead, viruses are classified as either DNA viruses or RNA viruses, depending upon the type of nucleic acid that's enclosed within the capsid. Viruses will contain either DNA or RNA, but never both. (If they had both, they'd be "alive.") The biggest distinction between the two types of viruses is in how they alter or use the machinery of the host cell.

Once it's inside a host cell, a DNA virus sometimes will directly produce new RNA, which in turn makes more viral proteins. Or, the DNA virus may actually join up with the cell's DNA and use it to direct the production of new viruses.

RNA viruses act a little differently. Once inside a host cell, an RNA virus may make new proteins directly. The virus releases the RNA, which migrates directly into the cell's cytoplasm and then uses its ribosomes to make proteins. The polio virus, which is an RNA virus, does this.

Other RNA viruses are called retroviruses. Besides RNA, retroviruses also contain an enzyme called reverse transcriptase. Reverse transcriptase makes DNA from RNA. In normal cells, the DNA makes RNA, and then the RNA makes proteins. In retroviruses, the RNA makes DNA (with the aid of the reverse transcriptase). Then the DNA makes new RNA. This new RNA makes proteins that become part of the new viruses. AIDS is a retrovirus. Because of the complexity and quantity of the steps involved, a retrovirus is the most difficult type to control or conquer. The virus is usually one step (or several) ahead of the scientists.

Diversity and Reproduction of Viruses

Scientists first learned about virus reproduction by studying viruses that infect bacteria, which are called *bacteriophages*.

Phages are easy to study because their host cells, the bacteria, multiply quickly in cell cultures, and scientists can see what's going on. Plus, if you recall, bacteria are prokaryotes; they're simple, and there's not a lot to keep track of. The phages most commonly studied are called T phages. T phages infect *Escherichia coli*, the bacteria that inhabit the human digestive tract. Scientists assign numbers to T phages, such as T1, T2, T3, and so on. The even-numbered T phages are virulent and capable of destroying *E. coli*.

Phages are one of the more bizarre-looking viruses around. They look like a sinister creature from a sci-fi movie because they have a polyhedral "head" and a "body" that is somewhat helical in appearance. They also appear to stand on "legs," which are called tail fibers.

Bio Buzz

A **bacteriophage** is a virus that infects bacteria. Scientists refer to these as phages.

Viruses reproduce through the lytic cycle. The *lytic cycle* is a fundamental process that occurs in all viruses. Because scientists use phages to study virus behavior, we'll use the T4 bacteriophage as an example of how this process works.

This T4 bacteriophage has a polyhedral head, as well as other viral features.

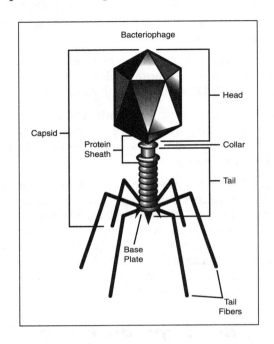

The lytic cycle has five sequential phases, including absorption, entry, replication, assembly, and release.

1. **Absorption**—During absorption, the virus attaches itself to the host cell. The tail fibers of the virus actually have a chemical affinity with the host cell's wall. The virus will attach only to specific spots on the cell wall, called receptor sites.

2. **Entry**—During the entry phase, the virus releases an enzyme that weakens the host cell's wall, or cell membrane. Then, much like a hypodermic needle, the virus presses its sheath against the cell and injects its own DNA into the cell through the weakened spot. With some

Bio Buzz

The term *lyse* means "to break open." The phrase **lytic cycle** refers to the breaking out of all the new viruses from the host cell.

viruses, the empty capsid remains on the outside of the cell wall. Other viruses enter their host cell intact. Once inside, the capsid dissolves and the genetic material is released in a process called uncoating.

3. **Replication**—During replication, the viral DNA takes control of all cellular activity. The genes contained in the DNA of the viral genome direct the cell to make viral DNA and viral proteins.

4. **Assembly**—During assembly, all the cellular activity is directed toward assembling the new viruses. The result is a host cell that is stuffed full of new viruses.

5. **Release**—During the release phase, the virus releases an enzyme that digests the host cell's wall from the inside out. This disintegration of the host cell, called lysis, releases all the new viruses. These new viruses go out and infect new cells, and the process starts all over again.

The T4 bacteriophage is a virulent virus. It enters the cell, takes over its machinery to make new viruses, and destroys the cell all in one continuous motion. But some temperate viruses invade cells without causing immediate destruction.

Scientists have studied temperate viruses the same way they've studied virulent ones, by using bacteriophages. Temperate viruses go through what is called the lysogenic cycle, which is actually a kind of life cycle.

The lysogenic cycle of a temperate virus starts much the same way as the lysic cycle: The virus attaches itself to the cell wall and injects its DNA into the host cell. But instead of immediately taking over the host cell's machinery, the new DNA just sits there. It becomes a component of the cell. If the host cell replicates and divides, the viral DNA goes right along for the ride but doesn't damage the cell.

The viral DNA really springs to life when the cell is exposed to external stimuli, such as radiation or chemicals, which kick the viral DNA into action. The virus then takes over the cellular machinery and destroys the cell.

Try It Yourself

Bacteria are everywhere, particularly on your own hands. Want to see just what's lurking there?

Take three pieces of bread without preservatives (preservative bread will work, but it will take longer). Rub your hands on one of the pieces when it's been a while since you washed your hands. Put the piece of bread in a sealed plastic bag, and label it. Then wash your hands carefully and rub the next piece of bread. Again, put it in a sealed plastic bag, and label it. Finally, take the last piece of bread and rub or drizzle it with some antibacterial soap; then rub the bread with your hands, put it in a sealed plastic bag, and label it.

Wait a few weeks and see what develops—you'll be surprised!

So, where did these things come from? How did viruses evolve? Unlike bacteria, there is no fossil evidence of viruses. But because viruses are parasites, scientists are pretty sure that viruses evolved after cells did. This is not a case of the chicken or the egg: The bacteria came first, and then the viruses showed up. Scientists think that viruses either formed spontaneously from nonliving organic matter (a bit like the primordial soup theory), or evolved from pre-existing cells.

However they got here, they certainly know how to evolve. Think about the common cold: Those viruses just invade and conquer. Your own immune system knocks out a lot of them, but not all. And those hardy survivors (which are now altered, thanks to battling your immune system) go right on to the next victim.

Think About It!

The common cold is a viral infection. If you've got a cold, there are virtually no drugs on the market that will help you—all you can do is wait it out. In the meantime, get plenty of rest, drink lots of fluids, and for reading material while you recuperate, try some biology books!

Bio Buzz

Morphology is the study of the internal and external appearance of an organism.

The short generation time of viruses means that natural selection acts with lightening speed and quickly selects the viruses most likely to withstand the next immune system attack. It's a never-ending battle.

Bacteria: The Basic Monerans

As discussed, the Kingdom Monera is composed of the least complex organisms. But what monerans lack in complexity, they more than make up for in their virulence and adaptability. "Disease" is frequently the first thing that comes to mind when you hear the word *bacteria*. And you're right: Bacteria do cause disease, some quite nasty stuff, like bubonic plague, typhus, and gangrene. But not all bacteria cause disease. Many monerans are decomposers, a vital link in the web of life. A decomposer is any organism that gets its nutrients from dead plants and animals. These are heterotrophs that eat only dead things. Decomposers are a valuable component in the web of life because if they didn't exist, nature would soon be buried in its own debris.

Types and Structure of Bacteria

All monerans are prokaryotic, which means that they are simple cells with no nucleus or membrane-bound organelles. They are also microscopic.

Monerans are considered to be the oldest living cells. Fossil monerans more than 3.5 billion years old have been found in Australia. From these simple beginnings, monerans have evolved into numerous forms and now live in nearly every environment, some of which cannot support any other type of living organism.

Within the Kingdom Monera, classifying bacteria into a hierarchy is difficult, mainly because they look so much alike. Instead, scientists use a mix of *morphology* and physiology when classifying bacteria.

Monerans come in different shapes, but most look like spheres, rods, or spirals. Spherical bacteria are called *cocci*. Rod-shaped bacteria are called *bacilli*. Spiral-shaped monerans are called *spirilli*.

Bacteria like to gather in groups. Some form filaments or chains, while others cluster. The prefix *staphylo* is used to describe bacteria cells that cluster, while the prefix *strepto* is used to describe bacteria that forms chains or filaments.

By combining the terms for bacterial structure and behavior, scientists have an accurate way to describe bacteria. For instance, a cluster of spherical monerans is called *Staphylococcus,* and a filament of rod-shaped bacteria is referred to as *Streptobacillus.*

Think About It!

Even the most inhospitable places on earth manage to support life. For more information, check out W. Hively's article, "Looking for Life in All the Wrong Places," in the May, 1997 issue of *Discover:* 76–85.

The Kingdom Monera is divided into four phyla, or group classifications: archaebacteria, schizophyta, cyanophyta, and prochlorophyta. Scientists think that present-day archaebacteria are the successors to the original primordial bacteria, the cells that first existed on Earth, primarily because archaebacteria inhabit environments where no other living organisms live. These environments are very similar to the original harsh conditions on early Earth.

Archaebacteria include methanogens (which live in the digestive tracts of sheep and cattle, and at the bottom of sewage treatment ponds, lakes, and bogs; they produce methane gas), halophiles (which live only in areas of high salt concentration, such as the Great Salt Lake and the Dead Sea), and thermoacidophiles (which thrive in very hot, acidic places, like in the geysers in Yellowstone National Park).

Schizophyta is the largest moneran phylum. As a phylum, Schizophyta breaks into four classes: Class Eubacteria, Class Actinomycota, Class Rickettsiae, and Class Spirochaeta. It's this phylum that's commonly referred to as "bacteria." It's sort of a catchall phylum, as all four classes are quite different from each other.

Eubacteria are free-living water and soil organisms, and they live in less harsh environments than archaebacteria.

Actinomycotes form into colonies of multicellular, branching filaments. Some actinomycotes cause diseases, such as tuberculosis and diphtheria. Some are decomposers

Think About It!

Bacteria living in your own home probably account for at least half of the 80 million cases of food poisoning each year. Take your kitchen sponges, for instance. Bacteria can live for two weeks on a wet sponge. When you use the sponge to wipe off counters and tables, you spread the bacteria. Antibacterial soaps, detergents, and bleach all kill bacteria, as does your dishwasher. To kill off kitchen bacteria, run your sponges through the dishwasher.

that break down dead plants and animals. Others are a source of antibiotics, particularly those of the species *Streptomyces*.

Rickettsiae can reproduce only in specific cells of a specific host. For instance, typhus is a rickettsiae bacterial disease that is transmitted by lice.

Spirochetes are spiral-shaped bacteria. Most of them have flagella that they use to move about. Syphilis and Lyme disease are both transmitted by Spirochaeta bacteria.

The third phyla of the Kingdom Monera is Cyanophyta. Cyanophytes are a unique type of prokaryotic cell, but they are also placed in the Kingdom Monera. These are often called blue-green bacteria and, like plants, can photosynthesis light.

Prochlorophyta make up the fourth phylum of the Kingdom Monera and are even more specialized than cyanophytes. They are photosynthetic bacteria that live symbiotically with a specific group of marine organisms known as tunicates.

You can see by all these variations that classifying an organism, particularly a microscopic one, isn't easy. In the lower kingdoms, it's particularly difficult to find unifying traits, especially among thousands of monerans, for instance. As scientists discover more about these organisms, the classifications sometimes change.

Nutrition and Nitrogen-Fixing

Monerans can be either autotrophic or heterotrophic. Most monerans are heterotrophs; they get their food from other organisms. A lot of monerans are saprophytes, which means that they feed on dead or decaying organic matter.

Monerans that produce their own food from inorganic matter are autotrophs, but within the Kingdom Monera, there are two types of autotrophs: photoautotrophs and chemoautotrophs.

Photoautotrophs have different pigments and chlorophyll that allow them to absorb light and, through photosynthetic processes, turn it into energy, or food.

Chemoautotrophs use the energy of chemical reactions to synthesize food. Some chemoautotrophs are also capable of nitrogen-fixing, which means that they can convert gaseous nitrogen into ammonia.

Nitrogen-fixing is important because plants need nitrogen to synthesize protein, but they can't use gaseous nitrogen. They can use only nitrogen that has been converted into ammonia compounds. Only monerans can convert nitrogen into a form that plants can use. Because all organisms ultimately depend on plants, all living organisms depend upon nitrogen-fixing monerans.

Pathogens

Any organism that causes disease is called a pathogen. Many known diseases are caused by pathogenic bacteria, including leprosy, tuberculosis, bubonic plague, and typhus.

Most pathogenic bacteria enter the human body through the respiratory system or gastrointestinal tract. In most cases, the disease is actually caused by *toxins* that are produced by the bacteria. Bacteria produce two types of toxins: endotoxins and exotoxins.

All endotoxins cause the same symptoms: weakness, fever, and damage to the circulatory system. Typhoid fever isn't actually caused by the bacteria, but rather by the endotoxins that the bacteria produces.

Exotoxins are metabolic products of the bacteria, secreted into the regions surrounded the bacteria. Exotoxins are the deadliest poisons known. Only a nanogram of the tetanus exotoxin will kill a guinea pig. Diphtheria and botulism are both caused by exotoxins.

Bio Buzz

A **toxin** is any substance that disrupts the metabolic activity of the infected organism.

Many pathogens can be destroyed by antibiotics. Unfortunately, however, *antibiotics* destroy all bacteria—good and bad—not just pathogens.

Many antibiotics are actually produced by living organisms. Penicillin was the first antibiotic ever discovered/invented, and it is a product of the fungus *Penicillium*.

One of the biggest problems with the use of antibiotics is that many bacteria become resistant to them through overexposure. When bacteria are first exposed to an antibiotic, most—but usually not all—of them die. The survivors frequently have genetic mutations that allow them to survive their next exposure to the antibiotic. They pass these traits on to the next generation of bacteria and the cycle of resistance builds.

Bio Buzz

An **antibiotic** is a chemical substance that can inhibit the growth of some bacteria.

The prevalent use of antibiotics has produced some particularly virulent strains of bacteria—bacteria that used to be, if not harmless, then responsive to standard antibiotics. Muppeteer Jim Henson's sudden death in 1991 shocked not only children and adults around the world; it also disturbed the medical community. Henson succumbed to a particularly virulent strain of group A streptococcus. This strain, which normally causes step throat, is now able to cause acute, fatal diseases in normally healthy people, sometimes within hours of the onset of symptoms.

Bacterial resistance to antibiotics doesn't come about just because you've had too many ear infections and taken too much penicillin in your lifetime (although that's certainly a contributing factor). Much of it comes from the widespread use of antibiotics in animal feed. It's the ol' heterotrophic chain at work: Not only do the animals consume the antibiotics, but every time you consume animal products, you do, too.

Many scientists believe that the overuse of antibiotics is applying tremendous selective pressure on bacterial strains, causing them to become much stronger than they normally would. Only time and scientific advancement will be able to answer this question.

We've covered quite a bit in this chapter: the five-kingdom method of classifying all living organisms, the unusual traits of viruses (which, thanks to their not-quite-alive characteristics, fall into a class by themselves), and the Kingdom Monera, which is the kingdom that's composed of bacteria. In the next chapter, we'll be talking about the most widely diverse kingdom of all, the Kingdom Protista, which includes organisms from algae to microscopic organisms that causes malaria.

The Least You Need to Know

➤ For purposes of classification, scientists divide all living organisms into five kingdoms: Monera, Protista, Fungi, Plantae, and Animalia.

➤ Kingdom classification is further broken down into phylum, class, order, family, genus, and specific epithet.

➤ Viruses bridge the two worlds of living and nonliving organisms. Viruses are alive, but not like other living organisms. A virus is a biological particle composed of genetic material and protein.

➤ Viruses reproduce by invading a host cell and taking over its machinery for their own uses, destroying the host cell in the process.

➤ Bacteria are the oldest, most primitive living organisms. Bacteria are the inhabitants of the Kingdom Monera.

The Catchall Kingdom: Protists

One of the purposes of the five-kingdom system is to allow scientists to classify organisms by their complexity. Starting with the Kingdom Monera, which has the most basic prokaryotic living organisms, the system progresses forward into more complex organisms.

The Kingdom Protista is a catchall kingdom. All its members are eukaryotic cellular organisms, meaning that they have a complex cellular structure with organelles, but other than that, they're quite diverse. Many organisms seem to be classed Protista, for lack of any other place to put them.

In this chapter, we're going to be talking about protists, the most assorted group of organisms on the planet. Because each type is so different from the next, (some swim, some ooze, some float, some snarf down food, and some engage in photosynthesis), they're really a great group to study. We'll be discussing the different types of protists that scientists know about (and they're pretty sure they don't know about all of them), their living habits, and accompanying dangers—some of these organisms are downright fatal to humans.

The Kingdom Protista

The organisms in the Kingdom Protista are more widely diverse than any other kingdom. The biggest thing the organisms have in common is that every protist is eukaryotic. In complexity, we've moved up a notch; the previous kingdom, Monera, was composed of prokaryotes, or very simple-celled organisms.

Scientists are sure that the first cells on Earth were prokaryotic. The oldest fossil cells ever found are about 3.5 billion years old and are prokaryotic. The first eukaryotic cells show up later in the fossil record and are about 1.45 billion years old. That doesn't mean that there weren't eukaryotes around before that—eukaryotes typically are soft-celled organisms, so they wouldn't leave a fossil. But the evidence indicates that eukaryotes evolved from prokaryotes in a process called *endosymbiosis*.

According to the theory of endosymbiosis, prokaryotic parasitic cells moved into other prokaryotic cells, probably to seek protection from the outside environment. Eventually, the prokaryotic parasites lost their ability to live outside the host prokaryote, and evolved into the cell organelles we see today inside eukaryotic cells. Scientists think that the mitochondria we see in animal cells actually evolved from parasitic bacteria. And they think that the chloroplasts we see in plant cells evolved from parasitic blue-green algae.

Bio Buzz

Endosymbiosis is the theory that prokaryotic intracellular parasites evolved into various cell organelles, forming eukaryotic cells in the process.

To encompass all these diverse organisms, the Kingdom Protista includes both plants and animals. But for clarity's sake, protists are generally broken into two rough categories: protozoa and algae.

Protozoans are animal-like protists. They're not necessarily what you think of as an "animal," but they're definitely not plants. All protozoans are heterotrophic, which means that they can't make their own food; some live freely and find food in the environment, and some live as parasites.

Free-living protozoans live in habitats where there is water, at least for some portion of the year. Some drift in water; others live in soil. Parasitic protozoans live in the bloodstream and tissues of their hosts. Most protozoans, free-living or parasitic, lack a protective outer coating, so the cell membrane acts as a semipermeable barrier to their environment.

Algae are plantlike protists. In stark contrast to their fellow protists, algae are gentle souls in that they don't bite or infect. In fact, they're sitting ducks when it comes to noxious chemicals.

All algae are autotrophic, which, as you remember, means that they can create their own energy. They form the foundation for the web of life and feed the planet. But aside from that, algae are a diverse group. Their methods of reproduction differ greatly, and in size, they range from microscopic to more than 100 meters tall. Nearly 25,000 species of algae are known. Now let's take a closer look at protozoa.

Protozoa

Scientists class protozoa into four different phyla: Sarcodina, Ciliophora, Zoomastigina, and Sporozoa. The main criterion for these classifications is method of movement. They're all eukaryotic, and they're all animal-like cellular organisms, but each phyla uses a different method of locomotion. You'll see when you read about each type next.

Phylum	Method of Movement
Sarcodina	Pseudopodia (cytoplasmic extensions)
Ciliophora	Cilia (hairlike extensions)
Zoomastigina	Flagella (whiplike tail)
Sporazoa	None, dependent on host or environment

Sarcodina: The Squishy Movers

About 40,000 identified species belong to the Phylum Sarcodina. Freshwater amebas are the most familiar sarcodines. Amebas feed on decaying organic matter found on the bottom of lakes and streams.

Many sarcodines live in freshwater and have flexible cell membranes (which, incidentally, is what enables them to move the way they do). Most of these sarcodines don't possess any added protective coating, although some possess calcium carbonate shells with spiky extensions. Still other marine sarcodines have silicon dioxide inside their shells. Because these little critters don't possess any appendages to help them move, they rely on a form of cytoplasmic streaming (if you recall, cytoplasm is the interior of a cell, what the organelles live in and what the plasma membrane contains) called *amoeboid movement* to get around.

The cytoplasm of a sarcodine is composed of two regions: the ectoplasm and the endoplasm. The ectoplasm is directly inside the cell membrane. Ectoplasm is somewhat like a thin layer of liquid right next to the cell membrane. The endoplasm is the rest of the cell interior, the cytoplasm and all the organelles.

Sarcodines move by pseudopodia, which are extensions of endoplasm that allow the protozoan to ooze along. To move, the endoplasm (enabled by the slippery ectoplasm) pushes outward, becoming distinguishable as a pseudopodium. At the same time the endoplasm pushes outward, previously formed pseudopodia draw back.

Through this pushing and retracting motion of its interior, sarcodines are able to move. It obviously doesn't offer much finesse, but it gets sarcodines where they need to go.

Nearly all freshwater sarcodines are similar to ameba. They absorb nutrients from the surrounding water through diffusion. They can also ingest food through phagocytosis, a form of ingestion through engulfing. When a sarcodine bumps into food, it surrounds the food with pseudopodia (that cytoplasmic streaming can by useful). Then the cell membrane pinches together and forms a food vacuole. Enzymes then digest the food.

Most freshwater organisms are sensitive to their environment because the water is always diffusing into them. To maintain a constant internal environment (homeostasis), most freshwater sarcodines have mechanisms that allow them to get rid of the excess water.

Ciliophora: Stroke! Stroke! Stroke!

Members of the Phylum Ciliophora propel themselves by means of cilia that line the outer wall of the cell membrane and move in synchronized strokes (kind of like a Viking ship).

Ciliates are found in both freshwater and marine environments. The best known ciliate is the freshwater paramecium. Paramecia (and other ciliates) are surrounded by a rigid protein shell called a pellicle. It's the pellicle that contains all the cilia—thousands of them, arranged in rows, beating in a wave motion. Because of this wave motion, as the ciliate moves forward, it also rotates.

Ciliates are internally different from sarcodines. The most distinctive difference between them is the presence of two nuclei in the ciliates: the macronucleus and the micronucleus. The macronucleus controls cellular functions such as asexual reproduction, digestion, respiration, and protein synthesis. The tinier micronucleus controls heredity and sexual reproduction.

Ciliates are well-adapted heterotrophs, and they have plenty of ways to feed on bacteria (or unlucky fellow protists that happen to come along). For example, paramecia have cilia-lined oral grooves. The beating cilia create water currents that sweep food into the orifice. From there, the mouth pore pushes the food into the gullet, which actually creates a food vacuole, or cavity. When full, the food vacuole breaks off from the end of the gullet and circulates through the cytoplasm, providing nourishment for the cell. The paramecium expels the leftover waste products through its anal pore. Ciliates reproduce through both asexual and sexual reproduction. Nearly all asexual reproduction occurs through the process of *binary fission*.

In most ciliates, sexual reproduction occurs through conjugation (and, no, it's not the same thing as conjugating a verb). In conjugation, individuals from two different mating strains (which means that they have two different sets of genetic material) line up next to each other, and in a fairly complicated process, exchange genetic material. With this fresh infusion of genetic material, ciliates then create new individuals.

Zoomastigina: Whip! Whip! Whip!

Members of the Phylum Zoomastigina (also referred to as the Phylum Mastigophora) are characterized by the presence of flagella. (If you recall, flagella are slender "tails" that extend from the organism. Some organisms only have one flagellum, others have several flagella.)The whiplike motion of the flagella pulls, pushes, and propels these protists through the water. Many zoomastiginoids live freely in ponds and lakes, where they feed upon other small organisms.

Other zoomastiginoids are parasites, capable of causing fatal diseases. Members of the genuses *Trypanosoma, Leishmania,* and *Giardia* are some examples of parasitic zoomastinginoids that can cause disease.

➤ **Trypanosomes** are slender, flattened protozoans with one flagellum. They live in the blood of various hosts (including humans) and are carried from host to host by blood-sucking carriers, such as flies. The best known disease caused by trypanosomes is African sleeping sickness. In this disease, the parasitic protozoans ultimately invade the host's brain (human), causing uncontrollable sleepiness, usually followed by death.

➤ **Zoomastiginoids,** of the genus *Leishmania,* are carried by sand flies. The disease they cause shows up in Africa, Asia, and Latin America. Leishmaniasis is characterized by skin sores and is frequently fatal.

➤ **Giardia** cause giardiasis, a disease characterized by diarrhea, cramps, fatigue, and weight loss. These particular protists are carried by muskrats and beavers. Humans play host to them when they drink contaminated water, frequently while on a camping trip.

Sporozoa: The Do-Nothing Protozoans

More than 6,000 species populate the Phylum Sporozoa, and all of them (at least in adult form) have no means of locomotion. All of them are parasites, and all of them

Think About It!

Disease can be found all over the world, but it's more rampant in some parts of the globe than in others. For a look at disease and how it is being combated on a global basis, check out the United Nations online, at http://www.un.org, as well as the World Health Organization (a division of the United Nations), at http://www.who.int/.

get around by traveling in the blood and bodily fluids of their hosts. They contribute nothing toward their own locomotion. And since they're parasites, you could also say they contribute nothing toward their own existence, relying totally on other organisms. Sporozoans obtain all their nutrition by absorbing it from whatever fluid they're floating in. Like many of their fellow protozoans, sporozoans are capable of causing disease in humans.

Toxoplasma and *Plasmodium* are just two sporozoans that affect humans. Toxoplasmosis rarely affects adults with healthy immune systems, but the disease is frequently fatal to newborns. One particular species of this genus also causes a deadly disease in young cattle and birds. Sporozoans of the genus *Plasmodium* are best known for the disease they cause called malaria. Malaria kills about two million people every year and is generally found in the tropics.

Plasmodium sporozoa have a very complex life cycle, which involves asexual and sexual reproduction, as well as more than one host. When a mosquito carrying the sporozoa bites a human, the *plasmodium* spores enter the person's bloodstream. From there, they go to the liver, where they reproduce asexually. These new spores infect erythrocytes (red blood corpuscles) and continue with asexual reproduction. Every few days, the new spores burst out of the erythrocytes and release toxins into the person's bloodstream. Some of the new spores develop into specialized cells.

All this is bad enough, but now the mosquito comes back into the act. When a female mosquito bites an infected person, it ingests the specialized cells. The cells develop into gametes—that is, sexual reproductive cells—that form into a zygote. A zygote is the first cell of a new individual. The nucleus of the zygote starts dividing, forming many, many internal spores. When the zygote membrane bursts, the new spores migrate to the saliva of the mosquito. And when the mosquito bites someone, the cycle starts all over again. (I know, the whole thing sounds like a far-fetched chain of coincidences. But with two million people dying every year, it's obviously not.)

For the infected person, the destruction of red blood corpuscles and the corresponding release of toxins into the bloodstream causes fever and anemia. Malaria also causes the spleen and the liver to become enlarged, and in acute cases, the kidneys fail. All this comes from a microscopic protozoan.

Algae: The Other Protists

Algae are the other members of the Kingdom Protista. They're the plantlike members (the gentle ones, the ones that produce oxygen for us and don't bite or infect anyone—we like algae).

For thousands of years, algae in the form of seaweed has sustained Asiatic cultures in a number of ways. In Japan, green seaweeds are used in soups and salads, red seaweeds are eaten fresh or toasted, and brown seaweeds are eaten raw or baked into breads. In North America, canned meat, salad dressing, toothpaste, sausage, chocolate milk, soup, and hand lotion, among other products, all have seaweed as a central ingredient.

Weird Science

Scientists used to classify algae as plants, but they no longer do so. Although all algae are autotrophic (they have chloroplasts and produce their own food and, indirectly, our food by photosynthesis), they do not reproduce by the same methods that plants do. Thus, they deserve a separate classification.

Algae are a diverse group of eukaryotic organisms, ranging in size from microscopic to more than 100 meters long. Yet all algae have several things in common. All algae cells have pyrenoids, organelles that both synthesize and store starches (which is one reason why they can be nutritious for humans—humans can metabolize starches for energy). Nearly all algae are aquatic, and those that aren't require water for reproduction. Most algae possess flagella as well.

Algae Structure and Classification

The body of any algae is called a thallus. The thallus comes in all sizes and shapes. Some are unicellular, some are colonial, some are filamentous, and some are thalloid.

Most unicellular algae are aquatic and are commonly referred to as plankton. Plankton and phytoplankton sustain our planet. Both provide food for marine organisms, and phytoplankton's photosynthetic abilities create much of the oxygen we breathe.

Weird Science

The word *plankton* is derived from the Greek word *planktos*, which means "to wander."

Colonial algae are independent algal cells, but they move and function as a single unit. Within the colony, a division of labor develops, with certain cells assuming certain duties. In this manner, the whole colony is able to move, feed, and reproduce more efficiently than if each cell was on its own.

Filamentous algae are algal cells that group in a linear pattern. By branching out, filamentous algae are able to anchor themselves to the ocean bottom and take advantage of environments that many other organisms can't reach.

Thalloid algae are organisms where the algal cells divide in all directions, creating a multicellular "thing" with leaflike, stemlike, or rootlike parts. Seaweed is a commonly recognized thalloid algae.

Scientists classify algae into six different phyla: Phylum Chlorophyta, Phylum Phaeophyta, Phylum Rhodophyta, Phylum Chrysophyta, and Phylum Pyrrophyta. Unlike protozoa phylum classifications, which are based solely on locomotion capabilities, algae are classified into a phylum based upon several things, with color being the primary determinant.

Members of each phylum have different colors. The color is determined by the photosynthetic pigments they have in their cells. All algae contain chlorophyll a, which absorbs more blue light. But different phyla contain additional chlorophylls, such as chlorophyll b, c, or d, each of which absorbs a different wavelength of light, thus giving the algae a distinct color.

Additionally, algae are classed according to their food-storage substances and the composition of their cell walls. Reproduction is another way to classify, although this criteria doesn't hold true across the board because algae have widely differing methods of reproduction.

Weird Science

Scientists think that green algae are the predecessors of plants for three reasons: First, both plants and green algae contain chlorophyll a and b. Second, both plants and green algae store their food as a starch. Third, both plants and green algae have cell walls that are made of cellulose.

Chlorophyta: The Green Algae

More than 7,000 different species make up the Phylum Chlorophyta. A green color is the defining characteristic of this phylum because its members can be unicellular, colonial, filamentous, thalloid, aquatic, or terrestrial.

Green algae are widely diverse. To understand a little about this group, it's probably best to look at some representative samples from each type.

Unicellular algae come in all varieties. *Chlamydomonas* is a common unicellular green algae that's found in freshwater streams and ponds, as well as soil. Its "body" basically consists of a single round-shaped chloroplast. Desmids, another type of unicellular algae, have thin, elongated shapes. Both, however, are unicellular green algae.

Colonial algae have all the cellular components of unicellular algae, but by grouping together in a colony, they also acquire some of the traits of a multicellular organism. A *Volvox* colony of green algae can contain up to 60,000 single cells. Taken as a whole, the colony exhibits an amazing division of labor, with different cells performing different functions, even though each cell is capable of performing any function on its own. The cells in a Volvox colony are able to chemically communicate with each other because they are connected by a fine web of cytoplasmic strands.

Filamentous algae tend to grow in long filaments. *Spirogyra* is one type of filamentous algae. It has unusual spiral-shaped chloroplasts that stretch from one end of the cell to the other. Other filamentous green algae, such as *Oedogonium*, have net-like chloroplasts.

Many types of seaweed are thalloid algae. Most are well adapted to their environment. *Ulva* has a leaflike body and can be found on rocks and pilings in the intertidal zone (the area between high and low tides). When the tide goes out, *Ulva* collapses its thalloid to prevent water loss. When the tide comes back in and *Ulva* is resubmerged, the thalloid plumps up again.

Phaeophyta: The Brown Algae

The Phylum Phaeophyta is composed of brown algae. There are more than 1,500 species of phaeophyta, and most of them are large multicellular marine organisms. The color of brown algae comes from fucoxanthin, an accessory pigment (in addition to the chlorophyll they possess) that is brown.

A typical member of the Phylum Phaeophyta is *Macrocystis*. This species of brown algae lives in the intertidal zone and can grow to be more than 100 meters long. The thallus, or algal body, is composed of a holdfast (the cellular part of algae that anchors it to rocks and other objects), a stipe (the stemlike part of the organism), and a blade (the algae's leaflike part).

The cell walls of brown algae contain a substance known as alginic acid, which is used to commercially produce alginates. Alginates are used to produce the gelling agents used in ice cream and other food products.

Rhodophyta: The Red Algae

More than 4,000 species belong to the Phylum Rhodophyta. A few species are unicellular and live in freshwater and terrestrial environments, but most rhodophyta are multicellular marine organisms. Rhodophyta are red algae and are rarely more than 1 meter long.

Red algae are unusual in that they can grow and survive in much deeper water than any other species of algae. They are frequently found at depths of up to 268 meters, with 150 meters being the most common depth. The "unusual" aspect of this comes from the fact that, like all other algae, red algae are photosynthetic. At those depths, how do they receive any light to photosynthesize?

In addition to chlorophyll a, red algae also contain chlorophyll d and some accessory pigments called phycobilins. Phycobilins are able to absorb the green, violet, and blue light that penetrates those depths.

The cell walls of red algae contain cellulose and a sticky substance called carageenan. Carageenan is used to produce, gelatin capsules, cosmetics, and some cheeses.

Chrysophyta: The Golden-Brown Algae

More than 10,000 species are a part of the Phylum Chrysophyta, and most of them are commonly referred to as diatoms. These are microscopic algae that float in ocean water.

Diatoms are an important source of food for marine life and are also responsible for the bulk of photosynthesis on our planet. This means that they are the foundation of life as we know it. They are an essential component of phytoplankton.

Diatoms are nonflagellated, photosynthetic algae whose shells are impregnated with silicon dioxide. They are unicellular or colonial and inhabit both marine and fresh-water environments. Diatoms have double walls and are highly ornamented and geo-metric. They look a bit like squared or rounded-off snowflakes.

Pyrrophyta: The Fire Algae

No, they don't shoot flames at unsuspecting scuba divers, but they do produce light (we'll talk about that in a minute). Members of the Phylum Pyrrophyta go by two dif-ferent names: fire algae and dinoflagellates. Most of the 1,100 species of dinoflagel-lates are marine and photosynthetic. Like chrysophytes, or diatoms, dinoflagellates are an important component of phytoplankton.

Think About It!

M. Satchell's article "The Cell From Hell," offers an interesting account of an algal bloom that killed 14 million fish. You can find it on pages 26–28 in the July 28, 1997 issue of *U. S. News and World Report.*

Dinoflagellates have cell walls made of cellulose and are shaped like tops—and they spin like tops as well. Their cellulose forms into plates, giving the cell walls the appearance of armor when seen under a microscope. Dinoflagellates have two flagella; when the flagella whip about, the organism spins like a top as it moves through the water.

Dinoflagellates have a unique characteristic: They have the ability to produce light (which is where the name "fire algae" really comes from). If you've ever seen the ocean glowing at night, chances are that what you were really seeing was a display of dinoflagellates.

Dinoflagellates are also responsible for a sometimes deadly phenomenon known as a red tide. Red tides are really population explosions of the algal species *Gonyaulax.* The water turns red because of the pig-ments in the algae. Red tides can be deadly because *Gonyaulax* produces a toxin. Mussels and other shellfish ingest the toxin from the surrounding water. If humans eat shellfish that have ingested *Gonyaulax* toxins, they can suffer a severe, sometimes fatal, neurotoxic reaction called mussel poisoning.

Euglena: The Misfit Algae-Protozoa

The nearly 1,000 species in the Phylum Euglenophyta are algae, but with a twist. Euglena are unicellular algae, and they have much in common with green algae, but they also have a lot in common with another species, the protozoa.

For instance, euglena contain chlorophylls a and b, and they store food as a starch. But they don't have cell walls, and they are not entirely autotrophic. If you place a euglenoid in the dark, for instance, it doesn't die; it becomes heterotrophic instead. When euglenas are in sunlight, they produce their own food through photosynthesis, but put them in the dark and they'll start eating anything they can catch. For euglenas, survival is the name of the game. Because of these characteristics, scientists generally classify euglena as protozoans because algae are not known to be carnivorous.

Euglena are abundant in freshwater lakes and ponds. They are long and fairly flexible. Like protozoa, they can change shape because they have a pellicle. They also possess contractile vacuoles—an organelle, which enables them to excrete excess water. Euglena possess eyespots, which guide them toward light for photosynthesis. A long flagella enables them to move through the water.

We've covered a lot of organisms in this chapter, both plants (algae) and animals (protozoans), some microscopic and some huge. The greatest commonality amongst the members of the Kingdom Protista is that they are all eukaryotic (complex-celled) organisms. We're moving up through the kingdoms, and from here on out, all the organisms we'll be talking about will be eukaryotes. In the next chapter, we'll be discussing fungi, nature's janitors and recyclers.

Think About It!

For your own personal taste of algae, visit any health food store or sushi bar. Nori, commonly sold by the sheet in packages, is a form of red algae. And those yummy California rolls? They're also wrapped in algae.

The Least You Need to Know

➤ The Kingdom Protista is the most widely diverse of all kingdoms. It is a catch-all kingdom for many types of eukaryotic-celled organisms. Scientists believe these eukaryotes evolved from prokaryotic cells.

➤ The Kingdom Protista is broken into two categories: heterotrophic animal-like protists called protozoa, and autotrophic plantlike protists called algae.

➤ All algae meet their own nutritional needs from sunlight. As a group, they photosynthesize the bulk of the oxygen on our planet.

➤ Euglena are protists that possess both protozoa and algae traits. They can produce energy from the sun but possess no cell walls. If placed in the dark, they meet their energy needs by eating other organisms.

Fungi!

You're probably noticing that as we move through the kingdoms of life, each kingdom gets a little more complex. Take fungi, for instance. These may not seem like complex organisms, but they're a lot more complex than bacteria, viruses, or protists.

The Kingdom Fungi is composed of such diverse organisms as yeasts used in baking, common mushrooms, and microscopic molds. Fungi can be both helpful and harmful. But regardless of which category they fall into, the main purpose of fungi is to recycle. If it weren't for fungi, most natural environments would become buried in their own garbage.

In this chapter, we'll be talking about the different types of fungi and what they do, what they look like, and how they reproduce.

The Main Event: Decomposing

Fungi are heterotrophs, which means that no fungi can produce its own food. They all survive by decomposing living or nonliving organic matter; that's how fungi get their

Bio Buzz

Fungi are heterotrophs that survive by decomposing living or nonliving organic matter.

A **decomposer** is any organism that obtains nutrients from breaking down dead plants and animals and absorbing them as food.

Think About It!

Fungi are a part of your everyday culinary world! You probably eat them fairly often because they include mushrooms, truffles, lemon drops, and soy sauce. All these things have fungal products in them.

nutrition. Unlike protists, few fungi engulf or catch their food (although there are some predatory fungi out there). To obtain nutrition, fungi spread out over organic matter and secrete enzymes, chemicals that break down the matter into "food." Then the fungal cells absorb the broken-down products.

Imagine the wooded areas of the north, brimming with life: deciduous trees and conifers, flowering plants, grasses, meadows and marshes. When the first winter storm blows through a forest, deciduous trees (oak, maple, beech, and similar varieties) drop leaves by the millions. In the space of a few hours, a 4-inch deep blanket of dead leaves covers the ground. Branches and twigs have snapped off the trees. Sometimes an entire tree has toppled over. Marsh grasses have flopped over and sunk into the mire. Meadows are stripped bare. By the time the storm passes, countless insects are dead, as are a few birds and some small mammals.

Winter storms are part of the natural cycle, and so is the death of plants and animals that result from the cold and dark. Go into any forest and start digging. As you work down through the accumulated layers of previous seasons, you'll find mucky leaves, remnant hard parts of insects and spiders, and crumbling branches. You'll probably even find the remains of a small mammal, such as a squirrel.

Every year, tons of debris drop onto the forest floor. A single elm tree can drop more than 400 pounds of leaves every autumn. At that rate, a forest would be buried in the space of a few years. Where does all this "garbage" go? It becomes nutrition for fungi. Fungi are nature's ultimate *decomposers* and recyclers.

Go back to the same forest in spring, and it's an entirely different picture. Trees are budding, flowering plants are pushing up, and the marsh grasses are springing to life. But that's not all. There's also a nearly invisible army afoot. Under leaves, on logs, along the banks, and under the water of the marsh, fungi are at work. All over the forest, puffballs, cottony molds, and mushrooms are popping up. Invisible to the naked eye, other fungi are at work.

Fungi do many things, but their main "purpose" in the web of life is to decompose and recycle nature's debris. Not only do they recycle dead plants and animals, but they're also known to break down photographic film, paint, leather, and clothing. They can even grow in jet fuel, clogging the fuel lines and causing all sorts of trouble.

Fungi tend to be either *saprobes* or *parasites*. Saprobes get their nutrients from nonliving organic matter, while parasites get their nutrients directly from a living host.

Whether saprophytic or parasitic, all fungi rely on extracellular digestion and absorption. Fungi are one of the few organisms that digest their dinner on the table, so to speak. Fungal cells secrete enzymes that are capable of breaking down large molecules into smaller components. The fungi then absorb these smaller molecular components.

In breaking down their food source, fungi alter it as they feed upon it. In decomposing, fungi make a lot of garbage "disappear." Their disappearing act provides food for the fungi and also recycles whatever the fungi doesn't use back into the environment.

When fungi break down garbage and debris, the process releases carbon and other nutrients back into the environment. Fungi don't really use these things, but plants do. So, essentially, decomposition by fungi produces the necessary materials for plants to produce food and oxygen for the rest of us.

Hing established that fungi are good and useful to humans, it's also necessary to point out that fungi are capable of causing some really nasty things: Athlete's foot, Dutch elm disease, and ringworm are just a few things caused by fungi.

Fungal Bodies

Most fungi have a multicellular, filamentous body that's perfectly adapted for absorbing nutrients and reproducing. The food-absorbing part of a fungus is called the *mycelium*. The mycelium is a mesh of tiny, branching filaments that spread out over (and into) organic matter, in all directions. This extensive network gives the fungus a great surface-to-volume ratio when it comes to absorbing nutrition. Each filament in the mycelium is called a *hypha* (plural, *hyphae*). Hyphae may or may not be divided into cells by a perforated cross-section called a *septum*.

Bio Buzz

A **saprobe** survives by getting its nutrients from nonliving organic matter. A **parasite** survives by getting its nutrients from a living host.

Weird Science

Fungi have even influenced the course of human history. Between 1845 and 1869, Ireland experienced several damp, cool growing seasons. This climate caused the rapid spread of *Phytophthora infestans*, a type of water mold that rots potato and tomato plants. At the time, potatoes were the main food source for much of the Irish population. Year after year, the fungus produced spores, causing rampant destruction. During that 15–year period, one-third of Ireland's population starved to death, died in the outbreak of typhoid fever (a secondary effect of the famine), or fled the country because of the effects of the fungi.

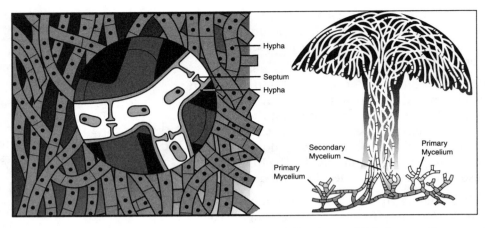

Fungi get their form from a branching network of filaments called the mycelium. Each individual filament in the mycelium is called a hypha. Hyphae may or may not be divided into cells by a perforated cross-section, called a septum.

Bio Buzz

Fungi are composed of a matted, branching, filamentous network called the **mycelium.** Each filament of the mycelium is called a **hypha.** Hyphae may be divided into cells by a perforated cross-section called a **septum.**

Hyphae are generally tube-shaped. The septum supports the hypha and helps give it shape. The perforations in the septum also allow the cytoplasm to extend continuously from one part of the mycelium to the other. Thanks to the hyphae and the septum, nutrients can move unimpeded throughout the entire fungal structure.

Sex Among the Fungi

In many fungi, the filamentous hyphae become interwoven, forming reproductive structures in the process. Within the filamentous mat, spores and gametes develop. Gametes are cells that are specialized for sexual reproduction.

The spores of filamentous species form with thick walls that enable them to withstand dry or cold spells (or both). When conditions are right, they germinate, giving rise to a new mycelium. Fungal spores are microscopic and durable, and can end up almost anywhere, thanks to the wind. The wind has managed to disperse black bread mold (*Rhizopus stolonifer*) all over the globe, including the North Pole.

Fungi are ancient, but not as old as some other organisms, such as bacteria. The oldest fossils that resemble fungi are about 900 million years old. The oldest fossils that are definitively fungi are about 500 million years old. By about 300 million years ago, all the modern divisions of fungi that we recognize today, had evolved.

Because life essentially started in the oceans, and because most present day fungi are terrestrial, scientists think that fungi went through an adaptive radiation shortly after plants and animals colonized dry land. Scientists also think that fungi (like other eukaryotes, or complex cells) arose from prokaryotic cells, the most simple of all cells.

Because all fungi are heterotrophic, which means that they can't make their own food, they probably arose from other heterotrophic cells. This isn't a universally accepted opinion, though. Some scientists think that some forms of fungi evolved from red algae, which are not heterotrophic. More than 70,000 species of fungi have been identified, although scientists speculate that there may actually be well more than 1 million different species.

Bio Buzz

A **fungal spore** is a walled cell or multicellular structure that is produced either sexually (through meiosis) or asexually (through mitosis). After formation, the spore is dispersed from the parent body.

Scientists classify members of the Kingdom Fungi into five phyla: Chytridiomycetes, Oomycetes, Zygomycetes, Ascomycetes, and Basidiomycetes.

The No-Fits: Fungus-Like Protists

Four phyla of fungus live in a never-never land between protists and fungi. For years, they were classified as fungi because their life cycle looked like that of a fungus and because they stored food as glycogen, again like a fungus.

But many scientists now consider these traits to be superficial, at least when it comes to classifying the organisms in question, and would like these phyla to be reclassified as protists, not fungi. Let's take a look.

Slime!

The Phylum Acrasiomycota contains about 65 species of cellular slime molds. (Sounds yummy, no?) Cellular slime molds live on land or in freshwater. The feeding stage of a cellular slime mold is a uninucleate cell called a myxameba. Myxamebas move and feed just like amebas, except that they do it on the forest floor, feeding on decaying plants.

Cellular slime molds are useful for physiological study because they're fairly simplistic. But because they are also eukaryotes, they're more complex than prokaryotes such as bacteria, which makes them useful for studying gene expression.

Plasmodial slime molds are placed in the Phylum Myxomycota. Plasmodial slime molds seem to be mostly cytoplasm. When they feed, the entire mold moves in a mass, feeding on organic matter as it goes. During times of stress, plasmodial slime molds become stationary and produce sporangia, or reproductive structures, on stalks.

In turn, the sporangia produce spores that can stay dormant for several years. When conditions are right, the spores release and, like magic, the plasmodial slime mold reappears.

Aquanauts and Others

Most fungi (and fungus-like protists) are terrestrial, but two phyla of the funguslike protists, Chytridiomycetes and Oomycetes, are found in water.

Weird Science

Both types of slime molds, cellular and plasmodial, exhibit amebalike movement, which is one reason many scientists think they should be classed as protists, not fungi.

Bio Buzz

The gamete-producing structures that develop within hypha are referred to as **gametangia** (singular, **gametangium**).

Most chytrids are saprophytic, which means that they obtain their nutrients by feeding on dead and decaying plants, like marsh grasses and reeds, found in aquatic environments. A few chytrids are parasitic. They obtain their nutrients from living plants and animals, as well as fungi. Motile spores (spores able to move) settle onto a host cell, germinate, and grow. At maturity, the fungal cell produces more spores. These swimming spores are released and go on to parasitize other cells.

Oomycetes, or water molds, are also funguslike protists. They appear to be only distantly related to other fungi. Oomycetes are the fungi that scientists think are related to red algae. Water molds are important decomposers of aquatic environments, and they can also be quite destructive. Water molds usually have an extensive mycelium. Gamete-producing structures, called *gametania*, also develop from the hypha.

Each gametangia is a separate reproductive structure. Fusion of the male and female gametes produces a diploid (two) zygote, which in turn, develops into a thick-walled spore. The spore rests until conditions are right and then germinates into a new mycelium. Water molds also produce asexual spores.

Besides being important aquatic decomposers, water molds are also indiscriminate. They decompose things without distinction. The fuzzy-looking growth that you sometimes see on tropical fish in aquariums is actually the mycelium of a parasitic fungus, *Saprolegnia*.

The ability of water mold spores (or any fungal spores, for that matter) to just hang around until conditions are right gives them a real advantage. When the fungus is destructive, this ever-present tendency can snowball with devastating consequences. A species of Oomycetes is what turned Ireland's potato blight in the 1800s into a famine. Because the weather conditions were so damp, because the spores were carried by water, *and* because the spores were capable of hanging around indefinitely, the famine and destruction encompassed the entire country for 15 years.

Zygomycetes: The Spoilers

Among scientists, there's no disagreement about the class of the members of the Phylum Zygomycota: They're fungi, and they're well known for their food-spoiling capabilities.

Most zygomycetes are terrestrial saprobes that live in the soil, feeding on decaying organisms. Many species inhabit dung and animal feces. Zygomycetes also feed on food that is being stored. Other zygomycetes are parasitic. These zygomycetes parasitize insects, including the common housefly.

Sac Fungi: Yeasts, Truffles, Morels, and Pathogens

Members of the Phylum Ascomycota are referred to as sac fungi, mostly because they produce haploid sexual spores in pouch-like sacs. They tend to be saprobic, feeding off of dead organic matter.

There are more than 30,000 species of sac fungi. The simplest members of the group are single-celled yeasts. Yeasts can occur naturally on fruit and leaves, as well as in the nectar of flowers. As you know, yeast also is another fungi that is commercially valuable to humans. *Saccharomyces cerevisiae* is a commercially manufactured yeast that produces the carbon dioxide that leavens bread. The species is also used to produce the ethanol in alcoholic beverages, such as beer and wine. In addition, yeasts are used in genetic research because they reproduce easily and quickly. Yeasts usually produce asexually, but many species can reproduce sexually.

Filamentous sac fungi are far more prevalent than yeasts. Sac fungi produce intricate saclike reproductive structures called *ascocarps*, which contain reproductive cells called *asci*.

Think About It!

To learn more about fungi, check out the gardening section of your bookstore. Indeed, fungi are the bane of gardeners—no gardener raves about the benefits of mildew. To find out more about fungi, take a look through some gardening books. There are any number of excellent books on the subject, including *The Rose Lover's Guide*, by Roland A. Browne (Atheneum, 1983). Yes, it's an older book, but after 300 million years, fungi hasn't changed much in the last 17 years!

Bio Buzz

Members of the Phylum Ascomycota are distinguished by their **ascocarps,** a saclike reproductive structure containing **asci,** or reproductive cells. Ascocarps are usually shaped like globes, flasks, or bowls.

Bio Buzz

A **lichen** is a symbiotic relationship between a fungus and green algae, or another photosynthetic partner.

Think About It!

The next time you see that long, straggly stuff hanging off trees in the Old South (where it's called "Spanish Moss") or the north woods (where it's called "Old Man's Beard"), you're actually looking at a lichen.

Truffles and morels are sac fungi. Both are edible and very expensive. Truffles grow underground, on the roots of trees, and are typically found in France, where trained pigs and dogs are used to sniff them out. Nowadays, truffles are cultivated commercially by inoculating the roots of trees with spores. Even so, they are still one of the most expensive luxury foods around, priced by the gram, not the pound.

Morels are found throughout the woods of North America and are prized by mushroom hunters everywhere. As with most edible fungi, morels can be misleading; There are true morels and then there are "false morels." The "false morels" are poisonous. (Forget Snow White and the apple; it's mushrooms you've got to watch out for.)

Several species of sac fungi are known for contaminating grain and wreaking havoc in bakeries. Dutch elm disease, chestnut blight, and apple scab are all caused by parasitic sac fungi. Pathogenic fungi are a double-edged sword. For instance, *Claviceps purpurea* is parasitic on rye and other grains, but its alkaloid byproducts also have medicinal uses, such as alleviating migraine headaches, returning the uterus to prepregnancy size, and preventing hemorrhaging after childbirth.

When its alkaloids are eaten in large amounts (which can happen when rye flour is contaminated by *Claviceps purpurea*), however, a disease called ergotism develops. Ergotism is characterized by halluinations, hysteria, vomiting, convulsions, diarrhea, and even gangrenous limbs. Severe cases are fatal.

Many sac fungi enter into symbiotic relationships with cyanobacteria or green algae. The term *lichen* refers to a symbiotic relationship between a fungus and its photosynthetic partner.

Lichens are often found in inhospitable places, such as bare rocks, rough tree trunks, and the Arctic tundra, to name a few. Lichen produce secretions that break down rock. In this regard, they are very useful, turning bare rock into soil that can support larger plants. In the Arctic, where plants are scarce, reindeer, musk ox, and other mammals can survive by eating lichen. In addition, lichens are useful indicators of air quality because they can't survive in heavily polluted air.

The symbiotic relationship of a lichen often begins when a fungal mycelium comes in contact with a free-living cyanobacterium, an algal cell, or both. The fungal mycelium parasitizes the photosynthetic cell, sometimes managing to kill it in the process. But if the photosynthetic cell survives, it reproduces and multiplies in partnership with the fungus. The fungus becomes so dependent on its photosynthetic partner that it can no longer survive without it. The once free-living fungus becomes entirely dependent on the captive algal cells for nutrition.

When it comes to food, the fungal cells are a bit of a hog. Often, the fungal half of the lichen consumes 80 percent of the carbohydrates produced by its photosynthetic partner. A few green algal cells actually benefit from a lichen relationship because they grow very slowly and can't compete well with other plants. These algal cells thrive in the shelter of a lichen relationship, but most photosynthesizers don't. The drain of their nutrients stunts their growth and inhibits reproduction.

Scientists classify lichens into groups according to their body shape, or thallus. Crustose lichens grow in a layer on rocks and trees. Foliose lichens are loosely attached and grow in leafy patterns. Shrubby lichens grow upright and are sometimes called two-fold lichens.

Club Fungi: Mushrooms and Puffballs

More than 25,000 species belong to the Phylum Basidiomycota, and chances are, you're pretty familiar with a lot of them. These include puffballs, mushrooms, and toadstools. Shelf fungi, stinkhorns, coral fungi, bird's nest fungi, and rusts and smuts also fall into this phylum. As a group, they are referred to as club fungi.

Club fungi generate club-shaped, spore-producing cells called *basidia,* located on the outer surface. The *basidiocarp* is the portion of the fungus that is visible above ground. A basidiocarp is what we refer to as a "mushroom." About 10,000 species of club fungi produce basidiocarps, but they are the only fungi to do so; no other species of fungi produce mushrooms.

Bio Buzz

The club-shaped sexual spores found in the Phylum Basidiomycota are called **basidia** (singular, **basidium**). Basidia develop on a short-lived reproductive structure called a **basidiocarp.**

When people look at a mushroom, what they are really seeing is a short-lived basidiocarp, composed of a stalk and a cap. The basidia, or spore-producing cells, occur on sheets of gills that are found on the underside of the cap. The rest of the fungus forms an extensive mycelium that grows into the soil, or whatever medium the fungus is growing upon.

When a spore is dispersed, hopefully, it will land on a suitable site and germinate. After an extensive mycelium develops, mushrooms will form if the conditions are favorable. If two different, yet sexually compatible, strains grow next to each other, cytoplasmic fusion may occur, eventually forming a zygote.

Members of this phylum run the entire gamut of fungal types. Species of club fungi can be saprophytic decomposers, symbionts, edible gourmet treats, or pathogens. This phylum includes everything from the cultivated mushrooms you buy at the grocery store to the smut fungi that causes serious plant diseases. Some club fungi enter into symbiotic relationships with photosynthesizers and form lichens. The group also includes hallucinogenic mushrooms used as recreational drugs mainly during the 1960s.

Try It Yourself

Put on your thinking cap, and see if you can advance the scientific cause by pondering these peculiar events:

One day you discover a fungus (*Trichoderma*) that is growing in distilled water. Thinking that there must be something else in the beaker, you vigorously scrub it down, making sure there are no traces of organic matter. But the fungus (which is not an autotroph) continues to grow, despite your best efforts to eradicate it. Why?

Long before penicillin or any other antibiotic was discovered, it was a common folklore practice to place moldy bread on a wound. Why would this help a wound to heal, if indeed it did?

Even if you're just looking for edible mushrooms to sauté up with your sirloin, they can be a tricky phylum to identify. In most cases, mushrooms look like mushrooms, so no general rules will help you when it comes to mushroom hunting. And mushrooms are unforgiving when it comes to mistakes. There are old mushroom hunters, and there are bold mushroom hunters. But there are no old, bold mushroom hunters. (Something to think about.)

Imperfect Fungi

Just as algae are primarily classed according to their color, fungi are primarily classed by one characteristic: their reproductive features. If the sexual phase of a fungi is missing (or, more likely, not yet discovered) the fungi is said to be "imperfect." When scientists discover a sexual phase of a fungi, it gets placed in the appropriate phylum classification, usually a recognized group, such as sac or club fungi. But until their sexual phase is discovered, several species of fungi sit waiting in the imperfect phylum. There are about 10,000 imperfect fungi species, otherwise known as the Phylum Deuteromycota.

Imperfect fungi can possess other traits that link them to a particular phylum; they are classed imperfect only because their sexual phase has not been completely determined.

And "imperfect" doesn't mean that the fungus isn't useful. Several imperfect fungi produce a type of sexual spore on their hyphae called a *conidium*. Many of these conidia-producing imperfect fungi are commercially valuable. *Aspergillus* produces the citric acid that is used to flavor lemon-flavored candies and soft drinks. It is also used to manufacture soy sauce; the fungus causes the soy beans to ferment into sauce. Certain species of *Penicillium* also produce the aroma and flavors found in Camembert and Roquefort cheeses. (The antibiotic penicillin is derived from a different species of *Penicillium*). Some imperfect fungi in the *Aspergillus* and *Penicillium* genuses have been reclassified into the Phylum Ascomycota because their sexual phase has been determined.

Other types of imperfect fungi are not so valuable, even causing many human diseases. The yeast *Candida albicans* is an imperfect fungus that causes yeast infections of the mouth and the vagina in humans. Other imperfect fungi grow in damp grain, and their toxic byproducts can cause cancer in humans who eat the poisoned grain over an extended period of time.

Fungi do more than decompose; they also form symbiotic relationships with plants. Unlike the symbiotic relationship of lichen (which darn near kills the photosynthetic partner), these other fungal symbiotic relationships, called *mycorrhizae*, benefit both parties. Mycorrhizae are typically found in forests. Fungal hyphae associate with the root system of coniferous and deciduous forest trees, as well as shrubs and other woodland plants. Most mycorrhizae are extremely beneficial to both parties. The fungus obtains carbohydrates from the plant; in turn, the plant absorbs ions from the fungus. The fungal hyphae are capable of extending out over a large distance, gathering mineral ions from a larger volume of soil than the plant's root system could reach on its own.

Bio Buzz

A **conidium** is an asexually reproduced spore found on the hyphae of Ascomycetes. Conidia are also found on some species of imperfect fungi.

Weird Science

Some imperfect fungi are predators. Hyphae of *Arthrobotrys dactyloides* actually form noose-like rings that snare unsuspecting nematodes (tiny worms). The hypha "noose" sits there until a nematode wanders in. The stimulation causes the hyphae to rapidly swell with water. The noose expands, and the hapless worm is caught. The hyphae then grow into the victim and digest it with enzymes.

105

Bio Buzz

A **mycorrhiza** is a symbiotic relationship in which fungal hyphae associate with plant roots. In most instances, the hyphae become densely entwined with the roots.

Mycorrhizae are extremely important to the health of a forest. In the absence of mycorrhizae, many plants cannot absorb mineral ions, particularly phosphorus, and whither and die—or at least fail to thrive. Mycorrhizae are also ecological indicators. They are extremely sensitive to acid rain, and this susceptibility is having repercussions in forests around the world. When the mycorrhizae die off, due to acid rain, the trees also die off—or at least fail to thrive.

We've covered a lot of ground in this chapter, particularly because we're moving into more complex organisms all around. We've talked about the primary role of fungi in the web of life (decomposing), as well as their structure, appearance, benefits, and disadvantages to humans. In the next chapter, we'll be talking about plants; the most basic component of life on Earth.

The Least You Need to Know

➤ Fungi are heterotrophs that are important decomposers within Earth's web of life.

➤ Fungi tend to be either saprobes or parasites. Saprobes get their nutrients from nonliving organic matter. Parasites get their nutrients directly from a living host.

➤ Fungi can be either terrestrial or aquatic.

➤ Although fungi spoil food and ruin crops, many fungi are beneficial. Penicillin is derived from a fungus, and many fungi, such as mushrooms are a valuable human food crop.

Chapter 9

Green, Green, Green: Plants

In This Chapter

➤ A bed of mosses

➤ Where the wild ferns grow

➤ Conifers and cones

➤ Flowering plants: more than just a pretty face

Plants are the most basic components of life on Earth. Through photosynthesis, they generate oxygen and form the base of the food chain that all life on Earth ultimately relies upon.

But plants do more than just supply food and air. More than 350,000 species of plants exist, and only about 1,000 of these are used for food. Humans also use plants for medicines, clothing, and building materials, to name just a few. Plants are found everywhere, even in the deserts of the world. Since the first algae managed to establish itself in the mud, plants have been colonizing the globe, inch by inch.

In this chapter, we'll be talking about the basic types of plants found on Earth. All plants can be classed as vascular (meaning that they have tissues that conduct water and offer support) or nonvascular (meaning that they don't have conducting and supporting tissues). Other than that, plants are a diverse lot. From barely visible "leaves" that float on ponds, to towering redwoods, there really is no such thing as a typical plant. Read on to see what we mean!

The Leap from Water to Land

About 430 million years ago, the only organisms found on Earth lived in the oceans. Mollusks and fish were diversifying, marine algae were flourishing, and coral were spreading out in warm, shallow waters.

But the Earth itself was a different story. Nothing existed there. The land was as hard, barren, and pitted as the moon.

Toward the end of the Silurian geological period, about 395 million years ago, tiny club-shaped plants began to grow in the shoreline mud. These plants were no bigger than your little finger, but they were plants. The first algae had made the jump from being aquatic to being terrestrial.

Weird Science

Scientists believe that all plants evolved from green algae of the Phylum Chlorophyta because both plants and green algae possess the same photosynthetic pigments (chlorophyll a and b), have cellulose cell walls, and store food as a starch.

Bio Buzz

A plant **cuticle** is a waxy, protective coating on the outside of the plant. The cuticle prevents desiccation.

Moving from the water to the land gave plants some distinct advantages over their oceanic counterparts. More sunlight was available for photosynthesis, more carbon dioxide was available, and there was less chance of becoming something's lunch (by this time, the oceans were teeming with animal life—"predators," as far as plants were concerned—and small photosynthetic organisms were getting scarfed on a routine basis).

But while land offered some real advantages, it also presented some challenges. For one thing, early plants were subject to drying out, or desiccation, through evaporation. One early adaptation these newly terrestrial organisms underwent was the development of a *cuticle*.

Plants that developed cuticles went on to thrive and reproduce, passing on the cuticle trait to their offspring. That was a good thing, because plants without a cuticle simply didn't survive. A cuticle was a necessary genetic mutation for early survival.

Gametes (reproductive cells like eggs and sperm) were also subject to desiccation. To overcome this problem and ensure successful reproduction, early land plants developed multicellular structures that protected their gametes during development. Protective gamete "holders," called gametangia, are a distinguishing feature of all modern plants—and they owe this trait to their algae ancestors that crawled onto the muddy shoreline. Again, it was survival of those best adapted. Once these basic protective features had been developed, plants started diversifying at a fast and furious pace. Some developed specialized structures that absorbed water from the Earth. Others developed complex tissues that transported water and nutrients throughout the plant. Supporting tissues also developed, which represented another necessary mutation because even a relatively small plant can be heavy. Without the buoyancy of water to support its weight, plants that moved onto land needed to develop supporting structures. This development of transporting and supporting tissues marked the development of vascular plants, because that's essentially what a vascular plant is: a plant with conducting and supporting tissues.

One of the biggest adaptations made by these early plants was the development of above-ground stems that absorbed light and underground stems that absorbed water and nutrients. Based on the fossil record, these light-absorbing stems are what evolved into true leaves. Underground stems evolved into true roots.

These first terrestrial plants reproduced through hardy spores that allowed for widespread distribution over land. Later, some plants developed seeds, which were a real reproductive advantage. Not only could seed remain dormant until conditions were right for reproduction, but they also offered a protect shell and nourishment for the embryo. Other plants managed to develop woody stems that allowed them to grow taller than other plants.

The Fossil Record of Plants

Most of what we know about early plants and the development of modern plants comes from the fossil record. The earliest known plant is about 400 million years old. This first plant, called *Cooksonia*, was only about 10 centimeters high and had forked, leafless stems. At the tip of each stem was the plant's reproductive structure. Because *Cooksonia* had vascular tissue (the type of tissue structure used for transporting nutrients and water within a plant) scientists are sure that it lived on land. Aquatic plants neither have nor need vascular tissues—they just absorb nutrients and don't need to transport them from the soil, nor do they need to support themselves against gravity; water does that for them.

Other leafless plants, including *Rhynia* and *Zosterophyllum*, show up in the fossil record at around 390 million years ago. And by the time the record reaches the Mid-Devonian period (about 345 million years ago), there are plants with true leaves.

Weird Science

The oldest living organism on Earth isn't an insect, a great blue whale, or even the descendant of a long-ago dinosaur family. It's an individual, one specific plant, a bristlecone pine tree that lives in California's White Mountains. If this tree possessed memory and voice, it would be old enough to tell us about the births of Jesus and Socrates. Called Methuselah, by scientists, its age is known, but other plants that are not yet accurately dated are also contenders. A particular box huckleberry bush may be 13,000 years old, and a certain creosote bush may be 11,700 years old. Scientists aren't 100 percent sure, though, because to definitively date any of these plants, they'd have to "kill" them.

Mosses and ferns dominated the Devonian period, with pine trees showing up in the Carboniferous period (about 280 million years ago). However, the dominant group of plants on Earth today, flowering plants, didn't show up until the Cretaceous period. Flowering plants are the most highly evolved of all plants and have been around for only about 135 million years, making them veritable babies.

Classification and Breakdown

Scientists divide plants into two basic groups, based on whether or not they are vascular, which means that they are vessels to transport internal fluids. Not only do vascular plants have water-conducting tissues, but most also have true stems, roots, and leaves. Nonvascular plants don't have vascular tissues, true leaves, roots, or stems.

Nonvascular plants have only one phylum: Bryophyta. With more than 14,000 species, mosses are the main nonvascular plants. In fact, mosses (and mosslike plants) are essentially the only nonvascular plants. They have no conducting or supporting tissues.

Vascular plants are further broken into two categories: seedless and seed. As you might suspect, these two categories really indicate how a particular plant reproduces; they either have seeds, or they don't. A seedless plant typically reproduces through spores, somewhat like the fungi we studied in Chapter 8, "Fungi!"

As we mentioned earlier, being able to reproduce through seeds is a real reproductive coup for a plant. Consequently, seeds are important—and not just for the plant. The type of seeds a plant produces and the structure of the seed itself represent an important method of classification for scientists. For instance, a pine tree and a maple tree are both vascular plants that produce seeds. But because their seeds are different, and because the method of reproduction is different, they're each in a different category. In this case, "a tree is a tree" doesn't apply.

Seed vascular plants are either gymnosperms (a category that includes conifers, pines, and other trees) or angiosperms (flowering plants and trees such as maples and oaks).

Mosses and Worts

There are about 23,000 species of nonvascular plants. Mosses, liverworts, and hornworts comprise the Phylum Bryophyta, which is the entire group of nonvascular plants. All nonvascular plants are bryophtes, and all bryophytes are nonvascular plants.

Scientists think that although bryophytes are nonvascular, they arose from plants that had vascular tissues. Some bryophyte species have nonfunctioning vascular features, which is the main reason scientists believe they are descendants of vascular plants.

Try It Yourself

The plant kingdom is huge. To give yourself an idea of the variety, take a walk around your neighborhood, local park, or wilderness area—any area with plants and trees will do. Collect one leaf from as many different plants as you can find. Try to get a representation of the different types of leaf and plant patterns: moss, fern, conifer, flowering plant, whatever you can find. When you get home, spread them all out, and take a look. Even a short walk should yield amazing variety.

Nonvascular plants seldom grow very large. Without the support that a vascular system offers, most bryophytes are small and lie close to the ground (or on tree trunks), where they can absorb water and nutrients.

Bryophytes are found only in places where there is water because they need water to reproduce: The sperm must swim through water to reach the egg. Most nonvascular species are found in damp climates, but they also can be found in deserts and dry regions, reproducing whenever there is some moisture. They can survive in extreme climates just by finding a tiny protected area. Next time you're hiking in the desert, look down into a protected crack in the rocks, and you'll probably find some moss growing there.

Mosses

Mosses are the largest class of bryophytes. Almost every terrestrial environment is home to one or more of the 14,000 species of moss.

When you walk on a thick carpet of green moss in a forest, you're actually walking on thousands of moss *gametophytes*, cells that produce sexual reproductive cells. Each one of these gametophytes is anchored to the forest floor by a root-like structure called a *rhizoid*.

Think About It!

If you're hurt while hiking, just find some peat moss; it'll keep you clean and knit together until you reach help. Mosses in general are pretty useful. Native peoples used moss for diapers, and peat moss was used as emergency bandaging material during World War I. It's highly absorbent and possesses antiseptic qualities.

Rhizoids are not roots in that they don't conduct water or nutrients to other parts of the moss. They only anchor and secure the plant. Mosses also have leafy structures, only one cell thick, that designed to photosynthesize and absorb moisture. Most moss plants are less than 15 centimeters in length, but they typically range from 1 centimeter to 1 meter. Like all nonvascular plants, without a supporting vascular structure, mosses just don't get very big.

Periodically, mosses reproduce when the sperm from one gametophyte fertilizes the egg of another gametophyte (and remember, there has to be water for this to occur, which means that reproduction doesn't occur frequently for mosses living in dry regions). In turn, the zygote (the cell produced when the sperm and egg unite) produces a stalklike sporophyte (the structure that holds the developing zygote, which in this case will be spores because mosses reproduce through spores). The sporophyte then forms a capsule. When the capsule matures, it releases spores that develop into tiny green filaments—"baby" moss, if you will. Eventually, these green filaments mature into gametophytes, and the process starts all over again.

Mosses are pioneers, boldly venturing where other plants fear to go. In this regard, they provide an important ecological function: They move into barren areas and, by gradually breaking up the surface of rocks, create soil for other plant species to move in and establish. (Those rhizoids are a lot stronger than they look.)

Mosses are frequently one of the first species to establish in areas that have been destroyed by fire, human activity, and even volcano eruptions. Mosses often trigger the development of new ecological communities. They're particularly useful in devastated areas because the rhizoids of moss hold the soil in place and help prevent erosion.

The Worts: Liver and Horn

Two other kinds of bryophytes exist: liverworts and hornworts. Mosses, though, compose the majority of the bryophtes, or nonvascular plants.

Liverworts come in different shapes and sizes. Some have a flat, or thalloid, body with distinct upper and

Bio Buzz

A **gametophyte** is the multicellular structure found in plants and algae that produces gametes, or sexual reproductive cells. A **rhizoid** is a rootlike structure that attaches the gametophyte to rocks, soil, or tree bark.

Weird Science

Sphagnum is a moss that comprises the peat bogs found throughout Ireland and other parts of northern Europe and Asia. In many of these countries, peat is a major source of fuel, and peat "mining" threatens the existence of these bog communities. *Sphagnum* gives off an acid that slows decomposition. Scientists have even recovered the intact—and perfectly preserved—remains of humans and animals that have been buried in bogs for thousands of years.

lower surfaces. Others have thin, transparent leafy filaments arranged around a thin, stemlike structure. But like all nonvascular plants, liverworts lie close to the ground so that they can absorb water and nutrients.

Hornworts look a lot like liverworts, except that their sporangia are slightly different. Hornworts also have a characteristic that links them with algae: Each of their cells possesses one large chloroplast rather than several small ones, like other plants.

Ferns and Other Seedless Vasculars

Aside from algae and bryophytes, all other plants are vascular, reproducing either with or without seeds. Scientists are puzzled as to why vascular plants are so numerous on Earth, but they speculate that it's because the specializations that vascular plants exhibit allow them to move into environments that nonvascular plants can't.

A supporting structure makes all the difference. Nonvascular plants are stuck on the ground, receiving spotty sunshine at best (usually because a vascular plant is taking all the sunlight and creating shade). But vascular plants can rise above the fray and get an unimpeded source of sunlight. Because they can procure more sunshine, they can make food more easily, and their vascular system allows them to efficiently distribute that food to all parts of the plant. (So, they grow even bigger and take away even more of the poor bryophytes' sunlight.) They also don't have to rely solely on water for reproduction. In times of drought, seeds just sit there. The vascular plant may die, but their seeds will spring to life when conditions are right. Besides their above-ground structures (stems and leaves), vascular plants also have root systems that enable them to get water from the soil—they don't just rely upon water they find on the surface. In short, vascular plants have just a bit more control over their own destiny than do nonvascular plants.

The conducting and supporting tissues that set vascular plants apart, so it follows that these tissues would be important and distinct to scientists. Two types of tissue make up all vascular plant systems: the *xylem*, which carries water and minerals, and the *phloem*, which carries sugar from the leaves to other parts of the plant. It doesn't matter whether it's a fern, conifer, or flowering plant—all vascular plants have the same two tissue systems. In addition to transporting, the way these tissues are constructed also offers support to the plant.

Seedless vascular plants had their heyday during the Devonian period. At that time, about 340 million to 350 million years ago, ferns were everywhere, as evidenced by their extensive fossil record. Nowadays, ferns aren't as common as they were

Bio Buzz

Xylem carries water and minerals from the roots of the plant to its stems and leaves. **Phloem** carries the sugars made through photosynthesis from the leaves (where photosynthesis occurs) to the rest of the plant.

then: Today, there are four phyla of seedless vascular plants. Phylum Pterophyta, or ferns, are the most numerous, with 12,000 species. The other three phyla are fern allies. Phylum Psilphya is comprised of about 10 species of whisk ferns. Phylum Lycophyta has about 1,000 species of club mosses. And Phylum Sphenophyta has about 15 species of horsetails. But all seedless vascular plants are considered ferns or fern allies.

More Seedless Vascular Plants: Ferns

Ferns are seedless vascular plants, in that they don't reproduce through seeds (it's still spores for this category), but they've got conducting and supporting tissues. Ferns most likely originated at the end of the Silurian period, about 420 million years ago. During that time, ferns, giant club mosses, and horsetails covered the Earth. They flourished for approximately 60 million years, during the Carboniferous period, and then most species became extinct. The ferns, whisk ferns, club mosses, and horsetails we have today are descendants of these original seedless vascular plants.

Most ferns are found in the temperate and tropical regions of the globe, but they also grow above the Arctic Circle and can be found in desert regions. They exhibit the same wide diversity when it comes to size. Tiny floating species are less than 1 centimeter across, while tropical tree ferns can be 25 meters tall. Some of today's most popular houseplants are ferns.

Except for tropical tree ferns, the stems of most ferns grow underground. This underground stem is called a *rhizome*.

Unlike mosses, ferns produce actual leaves. Fern leaves start out as tiny coils called fiddleheads. The fiddleheads uncoil and develop into mature fronds, or leaves. Every fern frond begins as a fiddlehead, even the fronds of the giant tropical tree ferns. Ferns produce sporangia and spores on the underside of their fronds. Turn a frond of any fern over, and chances are that you'll see small brown spots on the underside. These small "spots" are actually clusters of sporangia. When the spores are mature, the sporangia bursts open and the spores disperse to germinate and form new ferns.

Weird Science

The coal we burn today comes from the fossil remains of these giant fern forests of the Carboniferous period. Ferns fire our electric plants and heat our homes.

Bio Buzz

A **rhizome** is the underground stem of a fern. It contains vascular tissues and produces roots that absorb minerals and water from the soil.

More Seedless Vasculars: The Fern Allies

As their name implies, whisk ferns look just like whisk brooms and not much like true ferns. This is appropriate because, unlike true ferns, they have no roots or leaves, and they produce sporangia on the ends of short branches, traits indicating that whisk ferns most closely resemble early primitive plants.

Whisk ferns are found in moist, tropical and subtropical regions such as Hawaii, Puerto Rico, Florida, Louisiana, and Texas. Whisk ferns are also a staple of florist shops all over the world because they provide long-lasting greenery for floral arrangements and are easily grown and obtainable.

Other fern allies include club mosses, spikemosses, and quillworts. About 350 million years ago, these plants (referred to as lycophytes) covered the Earth, and most of them were the size of present-day trees. The fossils of these ancient lycophytes are the source of most of our modern coal deposits. Nowadays, lycophytes grow to a relatively small size.

Club mosses are the most familiar lycophytes. Club mosses are tiny evergreen plants that look like miniature pine trees. They have lots of leaves and a tough cuticle that helps them retain water.

Spikemosses are also small, vascular plants. Quillworts, too, are small. They live a submerged life in lakes and ponds, and have tubular leaves that resemble a feather quill.

The Phylum Sphenophyta contains horsetails, but they're still considered to be a fern ally. Horsetails have a jointed stem and leaves that look like scales. They thrive in the tropics, as well as alongside ponds and streams. They also grow well in dry, sandy places, such alongside roadbeds and railroad tracks.

Now we get into the seed plant categories: gymnosperms and angiosperms. In an evolutionary sense, seeds are a real success story. The most successful vascular plants, in terms of sheer numbers and range of distribution, are the seed-bearing species.

Think About It!

In many parts of the world, particularly the northeastern part of North America, emerging ferns, called fiddleheads, are considered a delicacy. For an interesting side dish, pluck the furled ferns and sauté them in a little butter and salt—yum!

Weird Science

Native Americans and early settlers used horsetails as scrub brushes. The outer cells of horsetails feel gritty to the touch and contain silicon dioxide, a major ingredient of sand. Horsetails made excellent pot scrubbers, which gave rise to their nickname, "scouring rushes."

As discussed at the beginning of this chapter, seeds really increase a plant's chances of reproductive success. Inside each tough, protective seed coat is an embryo, as well as a food supply for that embryo. When conditions are too hot, too dry, too wet, or too cold (sounds like Goldilocks, doesn't it?), the seed remains inactive. But when conditions are "just right," the seed germinates and the embryo begins to grow into a young plant, or seedling. Seeds are what carried many plants through the hot, dry Mesozoic period. While their seedless colleagues died off, the seeds of the seeded species just sat there, waiting for the right conditions to materialize again.

Every seed starts out as an ovule. An ovule is a cellular structure that contains a female gametophyte and egg cell. The gametophyte/egg cell is surrounded by tissues and protective cell layers.

When an egg becomes fertilized (united with a sperm cell), it develops into an embryo sporophyte. When this happens, the outer tissue and cell layers develop into a hard, protective seed coat. Basically, a seed is the structure that results from the maturation of a fertilized ovule.

Scientists divide seed plants into two basic categories: *gymnosperms* and *angiosperms*. Gymnosperms produce naked seeds that are not enclosed within a fruit. A pine cone is a typical reproductive structure for this group. Angiosperms produce seeds that are enclosed in fruit. As you might guess, an apple tree is an angiosperm—but in this case, "fruit" doesn't necessarily mean something that humans eat. For instance, a rose is also an angiosperm.

The ancestors of seed plants, as we know them, first showed up during the Devonian period. Based on the fossil record, scientists think that the first seed plants were gymnosperms.

Bio Buzz

The word **gymnosperm** comes from the Greek words *gymnos*, meaning "naked," and *sperma*, which in this case means "seed." Perched on the outer surface of a plant's reproductive structure, gymnosperm seeds are, indeed, rather "naked." The word **angiosperm** also comes from two Greek words: *angeion*, meaning "vessel," and again, *sperma* (still meaning "seed"). The seeds of angiosperms are enclosed (unlike the poor "naked" gymnosperms).

Conifers and Other Gymnosperms

There are four types of vascular plants with naked seeds, or gymnosperms: cycadophytes, ginkgos, gnetophytes, and conifers. With approximately 550 different species, conifers are the largest group.

Conifers are woody plants with needles, or scalelike leaves. This group includes firs, spruces, junipers, cedars, pine trees, yews, and redwoods, to name a few.

All conifers produce cones. Each individual conifer produces both male and female cones. The smaller male cones, called strobili, usually are produced in clusters. After releasing clouds of dustlike pollen, they fall off the conifer. The pollen fertilizes the larger, more complex female cones, which then close up tightly. Inside this protective structure, the developing seeds mature and are then released.

Most conifers are evergreen, meaning that although they do shed leaves throughout the year, they retain enough leaves at all times to distinguish them from deciduous species, such as maples, oaks, or elms, which shed their leaves every year. Because of their methods of reproducing, protecting, and dispersing seeds, conifers were able to expand into many different land environments during the Mesozoic period, a time when seedless plants were dying off. Nowadays, conifers are found all over the globe and are the dominant plant type in many regions. Although some are found in the Southern Hemisphere, most conifers are found in the Northern Hemisphere, particularly at high altitudes. Today, conifers are the major source of lumber and paper products.

Weird Science

The tallest trees in the world are conifers. The giant redwoods along the Pacific Coast of northern California grow to be 110 meters high, or the height of a 30-story office building. The oldest trees, bristlecone pines, are also conifers found in California. One particular bristlecone pine is approximately 5,000 years old.

More Gymnosperms: The Cycads

At first glance, cycads look like palm trees, but they're not related to palms at all; palms are angiosperms. During the Mesozoic period, cycads flourished along with the dinosaurs. Cycads have huge cone-shaped strobili that produce either pollen or ovules. The pollen from the male cones is transferred to the ovules of the female cones through crawling insects or air currents. Approximately 100 species have survived into present times, but their range is restricted to the tropical and subtropical regions of the globe. In Asia, people use cycads as food. They make the trunks into a starchy flour and also eat the seeds, after making sure to wash and rinse away the poisonous alkaloids that cycads produce.

Gingkos

Although gingkos were a diverse group during the Mesozoic period, only one species survives today. Gingkos are deciduous trees, meaning that they

Weird Science

The one species of gingko that remains on Earth isn't even a true species from the Mesozoic. The gingkos on Earth today are a domesticated population. Several thousand years ago, the Chinese planted gingkos extensively around their Buddhist temples. The natural population, from which these "temple trees" were cultivated, seems to have disappeared, leaving only the domesticated Chinese gingkos in its place.

shed all their leaves every year. They grow well in cities because they can tolerate air pollution and are resistant to insects. That's why scientists are baffled by their near extinction: They seem to be hardier than other trees. Scientists are uncertain as to why only one species of gingko made it from the Mesozoic period to the present.

Still More Gymnosperms: The Gnetophytes

Gnetophytes are found in both moist tropical regions and hot, dry desert regions. The more than 70 species include both trees and leathery-looking leafy vines. Their geographical distribution is just about equal, with approximately 30 species of gnetophytes living in the tropics, and about 35 species living in the desert areas. Gnetophytes produce cones, like other gymnosperms, but their vascular system is more like an angiosperm's than a gymnosperm's.

The appearance of gnetophytes is widely diverse. Some look like trees, while others crawl along the ground. One of the most bizarre-looking gnetophytes is the *Welwitschia*, a shrubby plant found in the south and western deserts of Africa. A woody, disc-shaped stem produces cone-shaped strobili and leaves, but the plant never produces more than two striplike leaves, which just keep splitting repeatedly as the plant ages. The result is a plant that looks like a scruffy heap. The bulk of the plant is a huge taproot that penetrates deep into the Earth.

The Flowering Plants: Angiosperms

The last category of vascular plants with seeds is the angiosperms, which produce flowers. Of all the plant groups, angiosperms are the most successful. Their more than 235,000 species have dominated the Earth for more than 100 million years. Within the tropics, new species of angiosperms are discovered on an almost daily basis.

For diversity, angiosperms can't be beat. You'll find them growing in northern forests, alpine meadows, prairies, deserts, wetlands and marshes, freshwater, and the oceans. Angiosperms range in size from less than a millimeter to more than 100 meters tall. Most angiosperms are free-living (meaning that they exist on their own and don't parasitize other organisms) and are photosynthetic. But a few, like the ghost pipes found on the floor of deciduous forests in northern North America, are saprobes, which, if you recall, are organisms that feed on dead organic matter.

Angiosperms first appeared in the fossil record about 135 million years ago, at the beginning of the Cretaceous period. At that time, gymnosperms dominated the Earth. But within 35 million years, flowering plants probably outnumbered gymnosperms. Besides having seeds and more efficient vascular systems, angiosperms had an important success factor: insects.

Angiosperms were evolving at the same time insects were. The fossil record indicates that this mutual evolution, or mutual adaptation, of both flowering plants and insects occurred during the Cretaceous period. Many flowering plants (both Cretaceous and now) can reproduce only when insects transfer pollen from one plant to another. The insects visit the flowers to eat, but in doing so, they transfer pollen and aid pollination. By developing along with insects, angiosperms were able to leap to the top of the pack, reproductively speaking.

Angiosperms are divided into two phyla: Phylum Monocotyledoneae and Phylum Dicotyledoneae. Put an easier way, flowering plants are classed as either monocots or dicots. Several traits distinguish monocots from dicots.

1. Monocots have a single cotyledon, or seed leave, in their embryos. Dicots have two.

2. Monocot flowers have a three-part symmetry to them. Dicot flowers tend to occur in fours and fives.

3. The leaves of monocots have a parallel venation, or "vein" arrangement. The "veins" of dicot leaves are arranged in a net pattern.

Weird Science

Angiosperms have evolved some amazing ways to attract insects. Some flowering plants are pollinated only by moths. These plants produce scent only at night, when the moths are active. Others produce tiny amounts of nectar, making it necessary for insects to visit plant after plant just to get adequate food.

Monocots

Monocots are not as numerous as dicots, but there are still a lot of them on Earth. Approximately 90,000 species of monocot angiosperms currently exist on Earth. Monocots are a diverse group, including such plants as corn, lilies, onions, bamboo, orchids, and leeks.

Monocot grasses are also responsible for much of the world's food supply and can be found in all parts of the globe, from Africa to Asia, to Europe, to the Americas. Rice, wheat, oats, and rye are just a few of the monocots whose primary use is food. Most of these food-crop grasses have been domesticated by humans for their use.

The stalks of wild wheat, for instance, break easily in the wind, an adaptation that increases seed dispersal but hinders the gathering of seeds for food. Early people selected the seeds from wheat plants with stems that didn't break easily. By planting these seeds, people were able to grow and harvest grain that didn't fall off the stalk.

Dicots

With nearly 200,000 known species, the majority of flowering plants are dicots. Dicots are the most numerous and diverse group of plants adapted to life on land. Dicots include familiar plants such as shrubs, trees (except conifers), and *herbaceous* plants.

Bio Buzz

An **herbaceous plant** is an annual plant that is nonwoody, or doesn't have the strengthening wood structure of a tree, for instance.

Maples, cacti, magnolias, and oaks, as well as daisies, roses, and oregano, are all dicots. By the Cretaceous period, plants and insects weren't the only organisms flourishing on Earth. Animals, both invertebrates and vertebrates, were also developing rapidly. The web of life was coming together.

We've covered a lot of basic material in this chapter, particularly because we've been talking about the foundation for the entire web of life. Given how numerous and important plants are, it makes sense that they'd be anything but simple. We've talked about the distinctions between vascular and nonvascular plants, as well as the distinctions between seedless and seeded plants. In the next chapter, we'll be discussing what all these plants help keep alive: animals.

The Least You Need to Know

➤ Scientists think that plants evolved from green algae and were the first organisms to make the leap from water to land.

➤ Plants are classed as vascular or nonvascular, based on their tissue structure. Nonvascular plants have no vascular, or water-conducting, tissues, nor do they have true roots, stems, or leaves. Vascular plants have water-conducting tissues, and most have true leaves, stems, and roots.

➤ Nonvascular plants are composed of mosses and worts.

➤ Vascular plants are further classed as seed or seedless. Seedless vascular plants are composed of ferns and their allies.

➤ Seeded vascular plants are further classed as gymnosperms (composed primarily of conifers) or angiosperms (composed of flowering plants).

Chapter 10

Flying, Crawling, and Other Creatures: Animals

In This Chapter

➤ Spineless creatures: invertebrates

➤ More than just oysters Rockefeller: mollusks and annelids

➤ Creepy crawlies: arthropods

➤ Stars of the sea: echinoderms

➤ Bones: chordates and vertebrates

Quick, think of an animal! Better yet, name 10 of them. If you were asked to quickly put down a list of 10 animals, it's likely that your list wouldn't include a single invertebrate (creatures without backbones), but you'd probably name a bunch of mammals, mostly furry critters with backbones.

Vertebrates are indeed animals, but they're a very small portion of all the animal species in the world. Most of the animal species on Earth are invertebrates. And in this case, "most" means 97 percent. Indeed, the animal kingdom is composed of organisms whose diversity ranges from sponges to humans, from lobsters to giraffes, from spiders to birds. They are all animals. In this chapter, we'll be talking briefly about all of them.

Classifying the Animal Kingdom

Within the Kingdom Animalia, invertebrates definitely rule, at least in terms of their sheer numbers. There are more than 1.5 million known species of animals in the world, and only about 50,000 of these are vertebrates, which gives you an idea just how skewed in favor of invertebrates the animal kingdom is.

Invertebrates cover a wide range of organisms: sponges, insects, worms, spiders, oysters, and others. Likewise, although vertebrates make up the minority in the Kingdom Animalia, they are also equally diverse, including goldfish, eels, turtles, manta rays, penguins, hippos, dogs, and humans. With so much diversity present within the animal kingdom, just how do scientists decide what organism is an animal and what's not?

For starters, to be classed as an animal, an organism must be a multicellular, motile heterotroph. It also must pass through some sort of embryonic development stage during its life cycle. Every animal will possess these traits, be it a barnacle or an elephant.

Invertebrates are a particularly diverse group because rather than sharing a common trait, or group of traits, they share the *lack* of a common trait: They have no backbone. Vertebrates share a common trait, and although they are a wildly diverse group, the presence of a backbone unifies them a bit more than invertebrates. Other than that, nearly anything goes.

Scientists use five body features to classify animals: body symmetry, cephalization, type of gut, type of body cavity, and segmentation.

Mirror, Mirror

An animal's body symmetry falls into one of two categories: radial or bilateral. The body of an animal with radial symmetry is arranged around a central axislike the spokes of a wheel. For instance, consider a squid. If you sliced its body lengthwise down the center, each half would look alike. And if you kept dividing the squid into equal sections (lengthwise down the center), like a pie, the sections would still look pretty much alike. That's because a squid's body exhibits radial symmetry.

But if you took a lobster and split it in half, those first two halves would look like mirror images of each other, but any further division would produce sections that looked nothing like each other. One piece would have a center back section, another would have only legs, another a piece of an antennae, and so on. That's because a lobster's body exhibits bilateral symmetry. The top surface of a bilateral animal is the dorsal surface, and the bottom surface of a bilateral animal is the ventral surface.

Sensation Is Everything

Scientists also classify animals based upon whether they exhibit cephalization. Cephalization is the concentration of sensory organs and nerve tissue at the anterior end, or head, of an animal. Both squids and lobsters exhibit cephalization; sponges and oysters don't. Cephalization is a real mystery to most scientists because it's hard to quantify sensation. Both you and a grasshopper exhibit cephalization, in that you both have a concentration of sensory organs and nerve tissues in the area called the head. Yet how do you quantify what either of you perceive and sense? If nothing else, cephalization gives you a good example of how complex the study of biology can be.

The Gut

The part of the body where food is actually digested and absorbed is called the gut, and scientists classify animals by the type of gut they have. Some animals, such as the sea star, have a saclike gut, which means that they consume food and expel wastes through the same opening. But in most animals, the gut is a complete digestive system, somewhat like a tube. Food enters through a hole (mouth) at one end, goes through several different functions as it travels the course of the tube, and is expelled through an opening at the opposite end of the tube (the anus).

Cavities, Anyone?

Animals also exhibit different types of body cavities. An animal's body cavity is the space between the gut and the animal's exterior body wall. Your body cavity is filled with lots of organs—there are a lot of *things* between your gut and your outer skin layer; it's not just empty space. But in other animals, the body cavity is essentially vacant, or else it's filled in with somewhat solid tissue.

Segmentation

Last (but not least), scientists group animals according to segmentation. Segmented animals consist of a series of body parts that may or may not be similar to each other. Earthworms, for instance, are segmented, and all their body segments look pretty much alike. Spiders are segmented, and their body parts (body vs. legs) don't look anything alike. In an evolutionary sense, segmentation offered animals the potential to develop specialized body parts and functions, and consequently move into specialized niches of the environment. Without segmentation, or the differentiation of specialized tissues from cells, we'd all still be simplex cells. Nearly every animal on the planet exhibits some type of segmentation.

Invertebrates Rule, Vertebrates Drool!

Invertebrates comprise most of the Kingdom Animalia. So far, there are at least 1.5 million known species of invertebrates, and that number rises consistently, mostly because a lot of invertebrates are small (think insects) and, consequently, are not easily discovered. Many are also marine organisms, and the ocean's a big place; scientists have yet to explore all of it. With so many diverse species in this group, the real defining characteristic among them all is the lack of a spine, or backbone. Other than that, nearly anything goes. Oyster, spider, earthworm, sea star—they're all invertebrates.

Primitive Invertebrates: Sponges and Cnidarians

The sponges you use to clean the kitchen sink are man-made, but they mimic the actions of a live sponge. They can soak up and release huge amounts of water because they're extremely porous—much more so than other animals. In fact, sponges are members of the Phylum Porifera, a name that comes from a Latin word meaning "pore-bearer." Divers continue to harvest sponges in the Aegean, Mediterranean, and Red seas, as well as the Caribbean, as they have since ancient times. Sponges have several unique qualities, including the following:

➤ Sponges are sessile, meaning that they attach themselves firmly to a surface and don't move about.

➤ Sponges neither chase nor capture their food. They filter whatever is floating in the water.

➤ Sponges grow in a wide variety of shapes, many of which resemble plants. They can look like a mossy mat, a cactus, or a blob of fungus.

➤ In size, they range from less than a centimeter in length to more than 2 meters in diameter.

➤ They also come in nearly every color of the rainbow.

Furthermore, the basic body pattern of a sponge is a hollow cylinder that's closed at the bottom, with an opening at the top known as the osculum. A sponge's body wall consists of two layers of cells separated by a jellylike substance. The outer layer is the pinnacoderm, and many scientists consider this to be the sponge's true tissue.

The cylinder itself is lined with special cells called choanocytes. The choanocytes have flagella; by beating the flagella, water is drawn into the sponge's body cavity through canals and pores that penetrate the sponge. (If you recall, flagella are long whip-like "tails.") As water passes through the sponge, its cells receive oxygen and food, and excrete wastes.

Without some sort of supporting structure, the body of a sponge would collapse in a heap. Some sponges have a simple skeleton composed of protein fibers. Others have a skeleton composed of hard little particles of silicon dioxide or calcium.

Sponges can reproduce sexually or asexually. Some species are either male or female, also known as *hermaphroditic*, but in most species of sponge, the individual produces both eggs and sperm. Sponges also have the ability to regenerate. A tiny piece of sponge can regenerate a completely new organism. In some species, particles small enough to pass through a cloth strainer will regenerate into a new individual.

Another primitive invertebrate is the cnidaria, better known by its common name: jellyfish. The phylum includes other organisms, such as freshwater hydra, flowerlike coral, and sea anemones. But stinging jellyfish—or sea jellies, as they are now frequently called—remain the best known cnidarians. Their very name derives from their cnidocytes, or stinging cells.

Like sponges, all cnidarians consist of two cell layers: the endoderm and the ectoderm. The layers are separated by a jellylike substance called mesoglea.

Bio Buzz

A **hermaphrodite** is any organism that produces both eggs and sperm. Self-fertilization rarely, if ever, occurs in hermaphroditic species. Sperm from one individual usually fertilizes the eggs from another individual.

Try It Yourself

Invertebrates come in all shapes and sizes. For a close look at one, go into any bath shop and buy a natural sponge. Take it home and examine it closely—maybe even cut off a small piece and look at it under a microscope. You're looking at one of the oldest animals on Earth.

Cnidarians use their tentacles to capture food. The tentacles capture prey and paralyze it with coiled stingers (called nematocysts) located in the cnidocytes. The tentacles then draw the prey into the mouth. Once inside the gastrovascular cavity (the hollow gut), enzymes break up the prey, and specialized cells lining the cavity absorb the nutrients. Waste products are expelled through the mouth.

Corals are small polyp cnidarians, and most of them live in colonies. Coral are unusual in that they secrete a calcium-based skeleton. It's these skeletons, fused together, that form coral reefs. Corals live only in warm, shallow, clear waters because they need sunlight to survive. Coral have a symbiotic relationship with algae, and these algae need sunlight to photosynthesize.

Unlike lichen, where the photosynthetic cell is often killed when the fungus takes it over, the partnership of coral and algae benefits both. The algae receives a safe home, and the coral polyp benefits from the oxygen the algae produces.

Weird Science

Portuguese men-of-war are one of the most unusual cnidarians. Each one exists as a colony of specialized polyps and medusae. Tentacles up to several meters long dangle from a gas-filled medusa, which can measure as much as 30 centimeters across. The tentacles of a Portuguese man-of-war have polyps that are specialized for feeding, digestion, and sexual reproduction. These giant jellyfish prey mostly on small fish, but their stingers, or nematocysts, contain a poison that's powerful enough to kill a human.

The algae are also responsible for much of the color seen in a coral reef. When scientists talk about the bleaching of a coral reef, they're talking about coral polyps under so much stress that they've expelled their algal partners. Stress can be caused by many factors, but the two largest factors are pollution and warming water temperatures. If the stress passes, the polyps will take in new algal partners and continue their symbiotic existence. But if the stress continues for any length of time—say, a year or so—the coral polyps will not accept any algae and consequently will die. Instead of a living, vibrant ecosystem, you've got a dead reef.

Weird Science

The largest structure on Earth ever built by living organisms was not constructed by humans—it was constructed by teeny invertebrates: coral polyps. The Great Barrier Reef in Australia stretches more than 1,200 miles and was built entirely by these little creatures. It's so large that it can be seen from outer space.

Worms, Worms, Worms: Flat and Round

As invertebrates evolved, their body patterns and structure got more complex. Sponges and cnidarians have some of the most simplistic body types. Worms—flatworms, roundworms, and rotifers—are more complex.

If you think back, the shape of an organism's body cavity is one criteria that scientists use for classification. This particular criteria, body cavity, can be broken down even further; scientists have identified four different types of body plans in invertebrates. The four different types are distinguished by the relationship between their *germ layers*.

The four invertebrate body types are defined by the presence or absence of the coelom, or body cavity. The four types go from least complex to most complex.

➤ **Two-layered acoelomate body plan—** This body plan has only two germ layers— endoderm and ectoderm—and they're not separated by a body cavity. Sponges and cnidarians have an acoelomate body plan.

➤ **Three-layered acoelomate body plan—**This body plan has three germ layers—endoderm, ectoderm, and mesoderm—but they're still not separated by a cavity. This body plan is a little more complex than the two-layered acoelomate plan. Flatworms have a three-layered acoelomate body plan.

➤ **Pseudocoelomate body plan—**Organisms with this body plan are characterized by a pseudocoelom, a cavity that forms between the mesoderm and the endoderm. Round-worms and rotifers are pseudocoelomates.

Bio Buzz

Germ layers are the layers of cells that originate in the developing embryo and become specific structures in the animal. Most animals have three germ layers: endoderm, ectoderm, and mesoderm. In animals, muscles, bones, and reproductive organs develop from the mesoderm.

➤ **Coelomate body plan—**This plan is the most complex and is characterized by the presence of a coelum, or true body cavity, that develops within the mesoderm. Mollusks, annelids, arthropods, and all echinoderms and chordates (including humans) are coelomates.

Because of this body plan distinction, "a worm is a worm is a worm" does not apply. They may look similar, but from a scientific standpoint, they're not. Some worms are more complex—hence, more evolved—than others.

Flatworms are bilaterally symmetrical animals that lack a circulatory or respiratory system. Many exhibit cephalization, in that their anterior region has an accumulation of sensory organs. So, although they look primitive, they can sort of tell where they're going—their anterior region enters a new environment first and transmits information to the rest of the body.

There are more than 13,000 species of flatworms. Not all of them exhibit cephalization, and not all of them even have mouths. But nearly all of them are parasitic. Humans play host to many of these parasites.

Roundworms and rotifers have a pseudocoelom body type. Their body cavity is lined with endoderm on the inside and mesoderm on the outside. The fluid-filled space between supports the body and is filled with organs. It also provides a storage area for eggs, sperm, and wastes, as well as the hydrostatic pressure that muscles contract against.

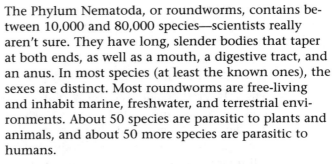

The Phylum Nematoda, or roundworms, contains between 10,000 and 80,000 species—scientists really aren't sure. They have long, slender bodies that taper at both ends, as well as a mouth, a digestive tract, and an anus. In most species (at least the known ones), the sexes are distinct. Most roundworms are free-living and inhabit marine, freshwater, and terrestrial environments. About 50 species are parasitic to plants and animals, and about 50 more species are parasitic to humans.

About 1,800 species populate the Phylum Rotifera. Rotifers are transparent, microscopic, free-swimming animals that live in marine and freshwater habitats. They have a crown of cilia around their mouths and feed on bacteria, algae, and protozoa. Rotifers maintain their shape by hydrostatic pressure. Their cephalization consists of two anterior eyespots and two long nerves that run the length of the body.

Bio Buzz

The word **mollusk** comes from a Latin term meaning "soft," referring to the animal's body. All mollusks have a soft body, although most also have a hard shell that protects (and frequently conceals) their soft body. Sea slugs and octopuses are some of the few mollusks that don't possess a hard shell.

Weird Science

The octopus is one cephalopod actually known for its intelligence. Octopuses can open jars to get food, go through mazes, and distinguish among objects of different size, color, and shape. They can even move from tank to tank when commanded to do so by a human—and can remember the experience. If that weren't enough, they can also regenerate severed tentacles.

Mollusks and Annelids

Mollusks consist of more than 100,000 widely diverse species. Within the Kingdom Animalia, only arthropods have more members. Although they don't look alike, clams, snails, and octopuses are all mollusks.

Mollusks exhibit diverse appearance, feeding habits, and methods of locomotion. Some mollusks are sedentary filter feeders, such as oysters and clams. Others are mobile and predatory, such as squids and octopuses. Diverse as they are, though, all mollusks have the following characteristics:

➤ A true coelom, or cavity between the gut and the body wall

➤ Three distinct body parts: the head, the muscular foot, and the visceral mass

➤ Organ systems for digestion, excretion, respiration, circulation, nerve impulse conduction, and reproduction

In addition, most mollusks have one or more shells and are bilaterally symmetrical. The head contains the sense organs, the mouth, and the cerebral ganglia. The muscular foot is probably the largest mollusk organ and is used for locomotion. The visceral mass contains most of the organ systems.

This third characteristic, the organ systems, really sets mollusks apart. Up until now, most of the organisms we've looked at have been relatively simplistic. It's a long journey from a single-celled bacteria to an organism with several organ systems.

Ironically, when it comes to mollusks, the fossil record is both rich and woefully lacking. We know plenty about the evolution of mollusk shells because shells are hard and preserve well. But the hows of soft mollusk body development remain something of a mystery because soft bodies are not well preserved or preserved at all.

Mollusks probably developed a hard shell of calcium carbonate early in their evolution. This had enormous adaptive value because the mollusks with hard shells were better protected. The development of a shell also caused problems, though—namely, it reduced the surface area that was available for gas exchange. In response to this problem, mollusks developed a related structural adaptation within the *mantle cavity*. These organs, called *gills*, allow the organism to exchange gases for water.

There are four major classes of mollusks, and scientists sometimes use up to seven to classify them. The four major phyla are Polyplacophora (chitons), Gastropoda (snails, conches, nudibranchs, abalones, and slugs), Bivalvia (clams, oysters, scallops, and other mollusks that have a muscular foot and a two-part shell), and Cephalopoda (squids, cuttlefish, octopuses, and chambered nautiluses).

Bio Buzz

Gills are organs specializing in the exchange of gases with water. In mollusks, the gills usually are located within the **mantle cavity,** the space between the **mantle** (the organ that secretes a shell) and the visceral mass.

Annelids

Scientists believe that mollusks and *annelids*, or segmented worms, share a common ancestor. Although at first glance a clam and an earthworm would appear to have little in common, they actually do. Both groups were probably the first organisms to develop a coelom, or a fluid-filled cavity within the mesosderm.

The strongest evidence for a common ancestor is found in both fossils and living species. Both mollusks and annelids share the same larval development. In both groups, the initial larval development is a pear-shaped larva called a trochophore.

Bio Buzz

The word **annelid** means "little rings" and refers to the numerous segments that make up the bodies of annelids.

With only about 15,000 known species, annelids are not particularly numerous, but they're certainly diverse. Marine feather-duster worms, blood-sucking leeches, and common earthworms are all annelids. Amid this diversity, annelids possess some common traits.

➤ All annelids have a true coelom, or body cavity.

➤ The bodies of annelids are divided into numerous segments. This phenomenon is referred to as metamerism.

➤ Most annelids have well-developed organ systems.

➤ Most annelids have external bristles on their bodies, called setae.

Weird Science

Believe it or not, you might run into leeches at your local hospital. Although the words *leeches* and *medicine* usually bring to mind a medieval sickroom, modern doctors still use them in combination with microsurgery, particularly in the reattachment of limbs and body parts. When microsurgery fails, it's usually because the soft tissues and tiny veins in the reattached part become clotted. When that happens, the tissues die and the body part can't rejoin the body and heal. But if a surgeon applies a leech to the reattached body part, the leech sucks out the congested blood. A period of prolonged localized bleeding follows. This gets the blood flowing correctly to the reattached part and really jump-starts the healing process.

There are three phyla for annelids: Phylum Oligochaeta (earthworms), Phylum Polychaeta (annelids living primarily in marine habitats), and Phylum Hirudinea (leeches, living primarily in calm, freshwater habitats).

Creepy Crawlies: Arthropods

Three-fourths of all the animal species in the world belong to the Phylum Arthropoda. This enormous phylum includes such diverse species as lobsters, crabs, millipedes, spiders, and insects.

Arthropods live in nearly every environment on Earth. Animals with arthropod characteristics showed up on Earth nearly 600 million years ago. They were here before the dinosaurs, and they're *still* here—and they'll probably outlast us all. Although they are a diverse group, there are some unifying traits among arthropods.

➤ **Jointed appendages**—An appendage is an extension of the body, such as an antennae or leg. The appendages of arthropods have joints.

➤ **Segmented bodies**—A pair of appendages is attached to each segment. In some species, evolution has diminished the size or removed the appendage.

➤ **Exoskeletons**—An exoskeleton is a hard, external covering that provides support and protection for the body.

➤ **Digestive systems**—Insects eat, digest, and pass their food.

➤ **Open circulatory systems**—Insects contain fluid (or blood) that transports gases and nutrients throughout their body.

➤ **Specialized sensory receptors**—Insects exhibit cephalization and other sensory organs, such as antennae.

➤ **Ventral nervous systems**—The nervous system of an insect extends to the outer parts of its body.

Because all arthropods possess common traits, scientists speculate that all of them evolved from a common ancestor. But current studies have suggested that there were most likely at least four different lines of arthropod evolution. Based on evolutionary characteristics, scientists class arthropods into four different groups: Trilobita (extinct trilobites), Crustacea (shrimp, crabs, lobsters, barnacles, crayfish, water fleas, and others), Chelicerata (spiders, ticks, mites, scorpions, horseshoe crabs, and sea spiders), and Uniramia.

Uniramia is the only group that seems to have evolved entirely on land. It includes millipedes, centipedes, and all insects. These insects account for the majority of arthropods. Although jointed appendages and a hard exoskeleton have certainly benefited insects (and all arthropods), it's their short life span that's really helped them. Because the adults of many species live for only a few weeks or months, generations occur in rapid succession, as does the process of natural selection. Any individual that has advantageous traits will survive and reproduce in rapid succession. So, any

Weird Science

Insects are a real biological success story, especially in terms of diversity and numbers. They live almost everywhere in the world, except the deep ocean. Water striders live on the surface of lakes and oceans, snowfleas live on glaciers at the top of the world, beetles live in the hottest deserts, aphids live on garden plants, and fleas live on your dog. More then 700,000 species of insects have been classified, and scientists speculate that as many as 10 million species of insects may exist.

Bio Buzz

Metamorphosis refers to the changes by which an organism passes through its life cycle into adulthood. An animal that goes through metamorphosis usually looks distinctly different during each stage.

131

Think About It!

While looking pretty, echinoderms can really cause problems, particularly in Australia. For an update on echinoderm problems, check out the Great Barrier Reef's Web site at www.gbrmpa.gov.au/cots.

Weird Science

Echinoderms can be voracious predators. One species in particular, the crown of thorns (COT), has been wreaking havoc on coral reefs in the Pacific and Indian oceans. These giant sea stars are 1 foot in diameter and love to eat coral polyps. When one arrives on a reef, it releases a chemical that attracts other COTs. This massive gathering then scours the reef, killing everything in their path. In the Maldive Islands, scuba divers are asked to pry off the COTs, in an effort to save the reef. At one resort, in only one year, divers removed more than 18,000 COTs. Because sea stars can regenerate, cutting up a COT only makes the problem worse.

evolutionary developments in insects progress rapidly because the turn-over time is quick, unlike that of, say, humans, who take decades to reproduce. Insects have also benefited from their ability to move into and survive in a wide variety of environments. They can do this because of their small size and their ability to fly.

Most insects also go through *metamorphosis*, and this, in turn, contributes to their success. Butterflies are a good example of an insect that goes through metamorphosis. They start out as eggs that become caterpillars. The caterpillars then spin a cocoon about themselves (which protects them), finally emerge as butterflies.

On the surface, metamorphosis wouldn't seem a contributing factor in the success of insects. But it eliminates competition between larva and adults (caterpillars eat leaves, but butterflies eat flower nectar), and the multistage life cycle helps insects survive harsh winters (most butterflies spend the winter as a pupa in a cocoon—they're not out flitting about, but they wait for warm weather and *then* they hatch).

Although insects are arthropods, they have four traits that distinguish them from other arthropods:

➤ Insect bodies have three parts: head, thorax, and abdomen.

➤ Insect heads have one pair of antennae.

➤ The thorax has three pairs of jointed legs. In many species, it also has one or two sets of wings.

➤ The abdomen is divided into 11 segments. An insect's abdomen has neither legs nor wings attached to it.

Echinoderms

Echinoderms form yet another group in the Kingdom Animalia. Echinoderms inhabit marine environments, ranging from shallow, sunny waters to the dark depths of 10,000 meters. Members of the Phylum Echinodermata include sea stars (formerly "starfish," but they're not fish), brittle stars, sand dollars, sea urchins, sea cucumbers, and others.

Adult echinoderms have an unusual body plan, in that it is radially symmetrical but also has some bilateral features. These creatures also move through a unique water vascular system that uses canals and tube feet. Most invertebrates swim, wiggle, or use their appendages to get around. Not echinoderms. Water enters their bodies through a sieve plate and flows into canals that run throughout their bodies. The pressure of the water creates suction in the thousands of tiny tube feet. The animal is able to move through this hydraulic pressure.

Echinoderms are definitely invertebrates, but they are classed by characteristics that few other invertebrates share:

➤ Most echinoderms undergo metamorphosis from a free-swimming bilateral larva to a bottom-dwelling, radially symmetrical adult.

➤ Most echinoderms have either five radii or multiples thereof.

➤ Echinoderms have an endoskeleton composed of calcium plates.

➤ Echinoderms have a water vascular system, or network of canals running throughout their body.

➤ Echinoderms have numerous tiny tube feet. These feet not only enable them to move, but they also aid in feeding, respiration, and excretion.

➤ Echinoderms have a nervous system, but no head or brain. They also have no circulatory, respiratory, or excretory systems.

Bio Buzz

A **chordate** is an animal that, at some stage of its development, has a notochord, nerve cord, and a muscular pharynx. A **notochord** is a dorsal chord of specialized cells, similar to a spine. The **nerve chord** resembles a hollow tube just above the notochord. The **muscular pharynx** serves as the entrance to the organism's digestive tract, and frequently its respiratory tract as well.

Chordates and Vertebrates

With *chordates* and vertebrates, we've reached the top, when it comes to complexity in the animal kingdom. We've arrived at the backbone of the Kingdom Animalia—literally. Compared to mollusks and arthropods, the Phylum Chordata is small, with only about 47,100 members. But this is the phylum with a recognizable creature in it: yourself.

All vertebrates and their close relatives are bilateral animals. There are three major groups of chordates: Urochordata (sea squirts and related forms), Cephalochordata (lancelets), and Vertebrata (fishes, amphibians, reptiles, birds, and mammals).

With only three general things defining a chordate, there's a lot of room for diversity. For most vertebrates, the *notochord* exists only in embryo form, quickly replaced by an endoskeleton. The notochord is not a true spine, particularly because there are invertebrate chordates—namely, sea squirts (or tunicates, as they're also called)—and lancelets. But it's a starting point because no other invertebrates have a notochord.

So, although sea squirts appear to be more closely aligned with a sea star or coral polyp, in an evolutionary sense, they have more in common with a fish.

Both tunicates and lancelets occupy separate branches of chordate classification. Tunicates have a notochord only during their larva stage. Lancelets retain the notochord throughout their life cycle. Yet both these animals form the evolutionary bridge between invertebrates and vertebrates. Who knows how many other "bridge" species there are in the world?

Vertebrates represent the highest evolutionary complexity. Eight different classes of vertebrates exist: Agnatha (jawless fishes), Placodermi (jawed, armored fishes—all extinct), Chondrichthyes (cartilaginous fishes), Osteichthyes (bony fishes), Amphibia (amphibians), Reptilia (reptiles), Aves (birds), and Mammalia (mammals). Of all the vertebrates, fish are the most numerous. More than 24,000 different species have been named, and no one has even attempted to figure out just how many billions of individuals this represents.

All vertebrates have highly organized, complex systems for dealing with gas exchange, circulation, digestion, reproduction, and just about any other bodily function you can think of. Attempting to understand all these systems comprises the bulk of any study of biology and life science.

We've discussed a lot in this chapter, mainly because animals are so diverse. We've talked briefly about all types of invertebrates (mollusks, worms, insects, crabs, and sea stars), the largest category of animals. We've also touched briefly on animals with backbones, chordates and vertebrates. In the next section of the book, we'll be talking about specific systems in animals. The next chapter will cover the skeletal-muscular system, mostly in humans.

The Least You Need to Know

➤ One way of classifying animals is according to their body plan. Animals are grouped according to the type of symmetry their body exhibits, their gut, their body cavity, whether they have a distinct head end, and whether the organism is divided into segments.

➤ Another way of classifying animals is based on the presence of a spine. Animals without a spine are called invertebrates. Animals with a spine are called vertebrates.

➤ The major groups of invertebrates, in order of complexity, are: sponges, cnidarians (jellyfishes and corals), flatworms and roundworms, mollusks (clams, octopuses, snails), annelids (segmented worms), arthropods (spiders, lobsters, insects), echinoderms (sea stars and sea urchins), and chordates (sea squirts and lancelets).

➤ All vertebrates have highly organized, complex systems for dealing with all aspects of being alive: circulation, gas exchange, nerve impulses, digestion, reproduction, and so on. Vertebrates are more biologically complex than invertebrates.

Part 3

How Animals (Human and Otherwise) Live

Do you live like a giraffe? How about a monkey? (Maybe your mother thinks you live like a pig, perhaps?) Animals exhibit enormous diversity, but they can still be recognized as animals, and although you may not like being compared to a pig or a horse, you're kind of stuck with it because all animals possess similar traits—particularly in how they manage to stay alive.

For starters, no animal can make its own food, so it has to forage for food. So you and that earthworm have at least that in common. But there's more than eating to being an animal. Skeletons, muscles, nervous and sensory systems, digestion, respiration, circulation, and more—all animals have these things.

When scientists group and classify animals, they're looking way past the surface, because that's not where the commonalities lie. They look at system organization. Anything that's alive is highly organized, and animals are no exception. Different species are arranged differently, they might look different from your own system, but every animal will have some variation of the same systems. In this part, we'll introduce you to you and your brethren in the animal kingdom.

Chapter 11

Movement! The Skeletal-Muscular System

In This Chapter

➤ Organization: always, always, always

➤ Skin: the all-purpose covering

➤ Dem bones, dem bones

➤ Muscles and how they work

The body of nearly every animal has an outer covering, muscle cells, and some form of skeleton. Each of these systems is complex and highly organized, no matter what the organism.

In humans, the three systems taken together form the image you hold of yourself, as well as how other people see you. Are you tall, short, dark, pale, hairy, bald, taut-skinned, flabby, fast, or slow? People may judge you on the surface, but your qualities are more than skin-deep (and in this instance, it's literal, not a cliche). All these qualities derive from your skeletal, muscular, and integumentary (skin) systems.

In this chapter, we'll be talking about your skeleton, the real structure of your body, as well as how your skeleton and muscles interact. And, of course, we'll talk about your skin, the largest organ of your body.

Always, Always, Always: Organization!

Think back to the qualities of life we discussed in Chapter 1, "Science and Life." One of the defining characteristics of life is a high degree of organization. Animals are alive, and they're definitely organized. Not only is the whole organism highly organized, but each system within the animal is also highly organized.

An animal's skin, bone, and muscles provide it with protection, support, and movement. No matter what the species, among individuals of that species, the organization—or layout, if you will—is going to be the same. All crabs will have the same muscle, skeletal, and integument systems. All grasshoppers will have the same muscle, skeletal, and integument systems. And all humans have the same muscle, skeletal, and integument systems, regardless of race. The organization of an animal's system will always be the same.

Skin Organization

All animals, from worms to humans, have an outer layer for their body. This outer layer is called the integumentary system. In most cases, an animal's *integument* is a tough, yet pliable, barrier between its truly interior systems (digestive, respiratory, and so on) and the environment. In crabs and insects, the integument is a hardened covering called a cuticle and consists of chitin, protein, and lipids. In vertebrates, the integument is skin.

So, in the simplest terms, your own integumentary system covers you. A human integumentary system is organized into two major components: the epidermis and the dermis. A third component, the hypodermis, anchors the skin and allows for freedom of movement, but it is not really part of your skin.

Bio Buzz

The word **integument** comes from the Latin word *integere*, which means "to cover." In humans, your skin covers your body, hence the name "integumentary system."

Skeletal Organization

All animals have some sort of skeletal system. Even sponges and sea anemones have a skeletal system, although it's not familiar or obvious to us. Three kinds of skeletal organization dominate the animal kingdom, including these:

1. **Hydrostatic skeletons**—The force of muscle contraction applied against internal body fluids transmits the force. Earth worms and sea anemones have hydrostatic skeletons.

2. **Exoskeletons**—The force of muscle contraction is applied against a rigid external body, such as a shell. Spiders, insects, and lobsters all have exoskeletons.

3. **Endoskeletons**—The force of muscle contraction is applied against rigid internal body structures, such as cartilage and bone. Humans and other vertebrates, such as fish, birds, and reptiles, have endoskeletal systems.

An animal's skeleton supports it and defines its body shape. Any skeletal system walks a fine line between offering support and remaining light enough for movement. In general, the larger the animal, the heavier its skeleton. As the size of an animal increases, the skeletal mass also increases to account for the disproportionate increase in the animal's weight.

Try It Yourself

You can play scientist right in your own kitchen. All you need is a chicken wing from your local grocery—or, if you're really ambitious, you can use a whole chicken. Chickens from the grocery store are a great way to learn about muscles and bones.

Clean your counter or work area well, and wash your hands (if you do so, you can turn your experiment into chicken soup when you're done). Get a small, very sharp paring knife, and you're ready to begin. (Some paper towels and forceps would be handy but not critical.)

Carefully remove the skin on the chicken wing, exposing the muscles and bone. If you pick up the wing and move it gently back and forth, you'll be able to see how the tendons, muscles, and bones interact. In one small, compact package, chicken wings contain all the essential ingredients. With further dissection, you should be able to observe how muscles are put together.

For instance, elephants have much heavier bones than deer. And whales, the largest vertebrates, have had to become (or remain, according to whichever evolutionary bent you're following) aquatic. If a whale's skeleton was strong enough to support its weight, it'd be so heavy that no muscle system could move it. A land-bound whale would have to become a plant, completely stationary. These weight vs. movement limitations apply to all animals, including insects, crabs, echinoderms, and vertebrates. Furthermore, human skeletons (and all vertebrate skeletons) are composed of bone and/or cartilage.

Muscular Organization

Without some sort of muscle structure, no animal could move. The force of a muscle's contraction against the skeleton (whatever type of skeleton the animal possesses) allows the animal to move.

In aquatic invertebrates, the muscle cells contract to bring in water (a.k.a. "nutrients") and also to expel wastes. Insects don't have muscles as we recognize them, but they have systems that interact with their skeletal systems and allow them to move. The muscle cells of most invertebrates, particularly those with hydrostatic skeletons, don't allow them to move with any sort of Olympic precision, but they operate and allow the animal to move.

The muscular system of humans (and all vertebrate animals) consists of three primary components.

1. **Smooth muscle**, which is found in internal organs, such as the stomach, and moves things through the body. Smooth muscles are involuntary, in that the animal can't control them.

2. **Cardiac muscle**, which is found in the heart. Cardiac muscle is also involuntary.

3. **Skeletal muscle**, which is what allows the animal to move. Skeletal muscle is voluntary, in that its movement can be controlled by the animal.

Weird Science

No garment has been made (nor probably ever will be made) that even begins to approach the qualities of what's already covering your own body. Only skin manages to maintain its shape despite repeated stretching and washing, is waterproof, kills many bacteria upon contact, screens the sun's harmful rays, repairs small burns and cuts on its own, and lasts as long as you do.

Skin: The All-Purpose Covering

All animals have what we call "skin." Among vertebrates, the integumentary system shows wild diversity. It might be decked out in hair, feathers, scales, claws, nails, horns, hooves, or quills.

Even among groups—fishes, for instance—diversity exists. Some fish have scales, while others have bare skin coated with slimy mucus. But no matter what it looks like, it's all part of the integumentary system, and it all serves the same purpose: protecting the animal's internal systems from the environment.

Skin protects your body from dehydration, bacterial attack, and abrasion. It helps control your body's internal temperature, and it has so many tiny blood vessels woven through it that it's also a reservoir for blood. These blood reservoirs are tapped during times of physical exertion, such as during strenuous exercise, and are shunted to other parts of the body, particularly your muscles. Your skin is also responsible for producing vitamin D, which your body needs to metabolize calcium. In addition, your skin is riddled with tiny sensory receptors that assess what's happening in the environment and then relay the information to your brain.

In general, human skin is only about as thick as a paper towel. The only places it's really thick is on your palms and soles, or in other places that are subject to abrasion and pounding.

The skin of vertebrates has two distinct regions: epidermis and dermis. For the purposes of this discussion, we'll concentrate specifically on human skin.

Think About It!

Don't ever take your skin for granted! It's actually your largest organ. An average-sized adult's skin weighs about 9 pounds. If you could stretch it out, your skin would have a surface area of 15 to 20 square feet.

The Epidermis

The epidermal layer of your skin is comprised primarily of stratified *epithelium.*

Each cell in your epidermis arises by itself but is pushed rapidly to the surface by the constant mitotic divisions of the cells underneath it. (If you recall, mitosis is the process whereby new cells are created by dividing in half.) Your epidermis is constantly losing old cells and producing new ones.

Most cells in the human epidermis are keratinocytes, so called because these cells are the primary producers of keratin, a tough, water-insoluble protein (your hair and nails are also composed of keratin, not just your epidermis). The exoskeleton and skin of many invertebrates also are composed of keratin. Keratinocytes start producing keratin when they reach the mid-epidermal region (remember,

Bio Buzz

Epithelial cells adhere closely to one another and are organized in layers. Epithelium always have one free layer, where no other cells can adhere to it. The epidermis of human skin is composed of epithelial cells.

your epidermal cells start at the bottom of the epidermal layer and are constantly being pushed to the surface). By the time they reach the surface, they're flattened and dead; all that remains are the tough fibers of keratin, packed into a plasma membrane. Millions of these keratin cells wear off daily, but thanks to rapid cell division and growth, new cells are always replacing them.

In your deepest epidermal layer, you find melanocytes, the cells that produce melanin and contribute to your skin color. Melanin is actually a brownish-black pigment that aids in protecting your body. The melanin gets transferred to the keratin-producing cells and provides a shield against ultraviolet radiation from the sun.

Skin color is also influenced by hemoglobin, the oxygen-carrying pigment found in human red blood cells. The red color of hemoglobin shows through the thin-walled blood cells as well as the epidermal cells, both of which are transparent.

The Dermis

Your dermis is made up of dense connective tissue, which serves as your primary defense against the countless environmental assaults launched against you every day. There are limits to what it can withstand, though, particularly if the assault comes from the inside. In pregnancy, for instance, the abdominal dermis undergoes such a stretching that it frequently ruptures, leaving behind visible white scars—"stretch marks," the bane of pregnant women. If a portion of your body experiences persistent abrasion, the epidermis separates from the dermis, which leaves you with a "blister." That's just another epidermis/dermis defense!

Most of your skin's components reside in the dermis: receptor endings of sensory nerves, blood vessels, lymph vessels, sweat glands, oil glands, and hair follicles, although some of these things actually derive from epidermal tissue.

Sweat glands secrete a fluid that is 99 percent water, along with some dissolved salts, ammonia, vitamin C, and waste products. You have about 2.5 million sweat glands in your body, and their primary function is to help control your temperature.

Oil glands are found throughout the dermis, in every part of your body except your palms and soles. The oil in these glands kills surface bacteria and lubricates both your hair and skin. If you have acne, the ducts of your oil glands have become infected with bacteria, and the glands have become inflamed.

All the hair on your body arises from hair follicles that are found in your dermis. A hair itself is composed primarily of keratinized cells. Each hair's root is embedded in the dermis, and only the shaft is visible. The average human scalp has about 100,000 hairs, but nutrition, genes, and hormones all influence hair density and growth.

Weird Science

All humans have approximately the same number of melanocytes in their epidermal layer. The variation in skin color that can be found in humans comes from the distribution and activity level of the melanocytes. African-Americans, for instance, have more active melanocytes in their epidermal layer than others. In addition, sun-tanning increases the activity level of melanocytes, which makes the skin look darker.

Think About It!

You don't have to wait for your own chronological aging process to cause your skin to deteriorate; you can do it on your own. Smoking, sun-tanning, and prolonged exposure to drying winds will all accelerate the aging process of your skin.

As you age, your epidermal cells divide less frequently, causing your skin to become thinner and more susceptible to injury. Glandular secretions slack off, so your skin becomes drier and less pliant. The collagen fibers in the dermis break down, and your skin loses its elasticity, so wrinkles form.

Dem Bones, Dem Bones: The Skeletal System

You skin is an excellent source of information. It's able to sample the environment and send information along to other parts of your body, such as your brain. Some of this information requires a response, namely movement. Your body moves through the activation, contraction, and relaxation of your muscle cells. But the muscle cells alone can't produce any movement. They need some sort of structure against which they can apply the force of the contraction. They need skeletons.

Although diverse, the three main types of skeletal systems (hydrostatic, exoskeleton, and endoskeleton) all have one thing in common: They provide the structure for an organism's muscle cells to contract against.

Hydrostatic skeletons, such as those found in sea anemones and earth worms, use fluid as the medium against which the muscles work. The containment of fluid in a confined and limited space supplies the pressure. Hydrostatic skeletons work somewhat like someone pushing on a full waterbed: The waterbed—or, in this case, the skeleton—supplies pressure and pushes back.

Invertebrates with hard exoskeletons lack the flexibility of soft-bodied animals, but they receive other benefits, particularly the ability to protect themselves from predators. An exoskeleton also provides support for an increased body size and the corresponding increase in weight, particularly when the animal is terrestrial and can't benefit from the buoyancy of water.

Exoskeletons afford more precision and speed of movement. All the hard parts of an animal's exoskeleton are hinged and attached to muscles

Weird Science

Animals with hydrostatic skeletons—which use fluid instead of bone as the medium against which muscles work—are able to move, but not with any kind of Olympic precision. Think about your waterbed trying to move—doesn't inspire confidence, does it? Animals with segmented bodies, though, such as earthworms, move a little better. Each segment has its own muscles and nerves, and is able to lengthen or shorten itself. So segmented animals, such as earthworms, are able to thrash back and forth, as well as move forward.

Bio Buzz

A **bone** is any part of the hard tissue that forms the skeleton of a vertebrate.

beneath the skeleton. An insect's cuticle covers the entire exoskeleton, even the gaps between segments, and provides some flexibility when the parts move.

Humans and other vertebrates are characterized by their endoskeletons. Some vertebrates have an endoskeleton of bone and cartilage, while others have an endoskeleton of cartilage alone. Sharks have an endoskeleton composed of cartilage that's hardened with calcium deposits. Most vertebrate skeletons, though, are composed primarily of *bone*.

Bones are complex organs, composed of a number of different tissues. These particular organs function in a number of ways:

1. **Movement**—By interacting with the muscle cells, bones change (or maintain) the position of body parts.

2. **Protection**—Bones provide a hard compartment for enclosing vital body organs such as the brain and the lungs.

3. **Support**—Bones provide a support system that anchors and supports the muscles and soft organs.

4. **Mineral storage**—Bone tissue is a repository for minerals that are vital for body functions. Calcium, phosphorus, and other minerals are all stored in bones. The body draws out and deposits these minerals as needed.

5. **Blood cell formation**—Bones are the sites of blood cell production.

Bio Buzz

Bone tissue consists of collagen fibers and living cells in a ground substance. Deposits of calcium salts harden the cells, fibers, and ground substance.

For this discussion, we'll concentrate on human bones (you'd rather hear about yourself than a fish, wouldn't you?), and specifically the long bones found in your limbs. Human bones range in size from tiny ear bones smaller than your fingernail, to clublike thigh bones. Bones are classified as either long, short (sometimes called "cubelike"), flat, or irregular.

Bones are made of tissues, including epithelium and connective tissues. Other organs contain these tissues, such as the epithelium found in your skin, but only bone contains *bone tissue*.

The thing that distinguishes bone tissue from other tissue is its extracellular matrix of ground substance and collagen fibers, both of which are mineralized.

Bone Structure

Consider your thigh bone. Compact bone tissue, which is very dense, forms each end of the bone. The density and hardness at each end of the bone allows it to withstand mechanical shocks. Most bones have denser tissue at their ends.

Your thigh bone, like other bones, is organized in concentric circles. The center of the bone is spongelike in appearance. Although it looks lacy and delicate, it's quite strong. In many bones, including your thigh bone, this area is filled with red marrow, which is where red blood cells are produced. Most mature bones have yellow marrow in this region. Yellow marrow is primarily fat, but it can convert to red marrow and produce red blood cells if your body experiences a severe blood loss.

The middle portion of the bone is a layer of compact bone tissue. It's composed of small concentric circles with canals of blood vessels and nerves running through the center of each circle, all of them compacted together.

The outer layer of bones is a thin, tough covering of dense connective tissues.

Bone Development

Long bones, such as your thigh bone, form during embryonic development. They start out as cartilage. Bone-forming cells, called osteoblasts, secrete a hard material inside the shaft region. This breaks down the cartilage, and the marrow cavity (which will become spongy bone tissue) forms.

Think About It!

Take care of your bones, especially as you get older. Bone development and turnover decrease with age, particularly among women. Osteoporosis is a condition in which the body keeps withdrawing minerals from the bones, without any corresponding deposits of new minerals. As more minerals are withdrawn, the bone tissue erodes and becomes hollow and porous, easily broken. Weight-bearing exercise, such as running or walking, is an excellent way to help prevent osteoporosis, as is actual weight-lifting (and you don't have to use heavy weights; studies in nursing homes have shown that even lifting soup cans produces an increase in bone density). Consuming at least 1200 milligrams a day of calcium also helps prevent bone loss.

The osteoblasts secrete the same substance on the outer surface, and this substance gradually turns the soft cartilage into hard bone tissue. The entire cartilage has turned into hard bone tissue, and in the process, the osteoblasts have become trapped inside. Once trapped, they are called osteocytes and contribute to the maintenance of mature bones.

Bones are constantly developing. Bone tissue turnover occurs when adult bones are subjected to exercise, stress, or injury. In your thigh bone, for instance, the diameter increases when bone cells deposit minerals on the shaft's surface. At the same time, other bone cells destroy an approximately corresponding amount of bone tissue within the shaft. So, your thigh bone becomes thicker and stronger, but not much heavier.

Minerals are constantly being withdrawn and deposited from bone tissue. Exercise stimulates the deposition of calcium, while stress or injury causes a withdrawal. This bone turnover is how your body maintains its calcium levels. When your body needs calcium, bone cells secrete enzymes that break down the bone tissue. The calcium (and other minerals, because your bones are more than just calcium) enter the interstitial fluid (the fluid that fills the spaces between cells) and are picked up by the blood. Osteoblasts are constantly active, even into maturity. The metabolic activity of these cells keeps your bones healthy and strong, and also allows your bones to repair damage.

Human Skeletal Structure

About three million years ago, humans distinguished themselves from most other vertebrates by walking upright on two legs, instead of four. Unfortunately, this isn't an ideal arrangement, at least when it comes to the design of the human skeleton. The problem is, standing upright places a great deal of stress on your back, thanks to the pull of gravity. The older you get, the longer your body has been resisting this pull and, usually, the more lower back pain you experience.

Your skeleton is divided into two major regions: the axial skeleton (which includes your skull, backbone, and ribs—essentially the central axis of your body), and the appendicular skeleton (which includes all the limbs and other bones in your body. Altogether, you have about 206 bones.

Joints are the areas of contact (or near-contact) between bones. Synovial joints are most familiar. These are the freely movable joints in your body, such as your knees and your elbows. Synovial joints are stabilized by ligaments (dense, connective tissue that bridges a joint).

Cartilaginous joints have cartilage between the bone spaces, which permits only slight movement. Cartilaginous joints are found at the junctures between your vertebrae.

Fibrous joints unite bones with fibrous tissue, and no cavity for movement is present. The bones in your skull start out as fibrous joints in infancy and childhood. But by the time you're an adult, these joints have fused into a single unit.

Muscles and How They Work

Your skeleton supports and protects your body, but without muscles, you'd be just another pretty mannequin. Remember that one of the characteristics of life is movement. Muscles enable animals to move.

The three types of human muscle (skeletal, cardiac, and smooth) are all quite different in appearance, but they have three traits in common.

1. All muscle cells exhibit excitability, in that they respond to stimulation.

2. All muscle cells can contract, or shorten, in response to stimulation.

3. All muscle cells are elastic. After contracting, they return to their original relaxed position.

Skeletal muscles are the only type of muscle that interacts with the skeleton to bring about changes in the position of body parts. They are the only muscles responsible for locomotion. Cardiac muscle keeps you alive, and smooth muscle helps you digest food, but neither contributes anything toward getting you up off the couch. In this section, we're going to talk only about skeletal muscle.

As you may remember, all cells have a plasma membrane. But what we haven't told you until now is that there is a voltage difference across the plasma membrane. The electrical charge on either side of the plasma membrane will be different. (Ever thought of yourself as electrical? It's somewhat like the electron transport chain described in Chapter 5, "Physi-What?") In excitable cells such as muscle cells, the voltage can reverse suddenly and briefly with adequate stimulation. This sudden reversal in the charge of the cell is called a nerve impulse (sometimes referred to as an action potential). All muscle cells have action potential.

This nerve impulse, or charge reversal, in its plasma membrane causes muscle cells to contract. Because the charge reversal is sudden and brief, and because all muscle cells are elastic, the muscle cells return to their relaxed position right after they contract.

Weird Science

Humans aren't the only vertebrates with muscles. All vertebrates and invertebrates have muscles. When a hummingbird hovers at a feeder, it's doing so solely on the strength of its skeletal wing muscles; no significant air force is helping it defy gravity. Bats and insects such as bees, dragonflies, moths, and butterflies can also hover, courtesy of their skeletal wing muscles.

Bio Buzz

A **myofibril** is the contracting threadlike structure of a skeletal muscle. Each myofibril is made up of thick and thin filaments, all side by side in a parallel pattern.

Skeletal Muscle Structure and Movement

Skeletal muscles are composed of anywhere from a few hundred to thousands of muscle cells. Strong connective tissue encapsulates the muscle cells and also forms the tough tendons that attach both ends of the muscle to the bone.

Each muscle cell is constructed of threadlike structures called *myofibrils*. Each "thin" filament in the myofibril is actually two beaded strands, twisted together. And each "bead" is actually a molecule of actin, which is a contractile protein. Each of the "thick" filaments are packed-together molecules of myosin, another contractile protein.

The actin and myosin filaments make up sarcomeres, which are the basic units of muscle contraction. Ever noticed the striped appearance of muscles? It comes from the sarcomeres. The organization of the actin and myosin filaments into sarcomeres is so highly ordered (a place for every filament, and every filament in its place) that it gives sarcomeres—and muscles—a striped appearance.

So how do sarcomeres contract? How do your muscles move? According to the *sliding-filament model*, the myosin filaments physically slide, pulling the actin filaments toward the center of a sarcomere during contraction.

The sliding movement requires the use of energy (remember ATP, the storage molecule for energy?) and the formation of cross-bridges between the myosin and actin filaments. The cross-bridges form when the head of a myosin molecule attaches to a binding site on the actin molecule. An ATP molecule is associated with each myosin molecule, and when its energy is released, the myosin "head" tilts forward (like a power golf stroke) toward the center of the sarcomere. As actin filaments become attached to the myosin, they get carried forward also. More ATP causes each myosin head to detach and reattach at the next actin site, again moving everything forward to the center of the sarcomere.

A whole series of myosin "power strokes" in each sarcomere are required for a single muscle contraction.

Controlling Those Contractions

Ultimately, your physical strength depends upon how forcefully your muscles can contract. (Arnold Schwarzenegger's muscles contract *very* forcefully). What causes these contractions?

Your motor neurons are what command your muscles to contract. These neurons stimulate action potentials to travel along the plasma membrane of a muscle cell. They continue along to the sarcoplasmic reticulum, a system of membranous chambers around the myofibrils. These chambers serve as a store house for calcium ions.

When the nerve impulses arrive, they cause the sarcoplasmic membrane to become permeable to calcium ions. And after they're released, the ions diffuse through the membrane and attach to the actin filaments. When the calcium attaches, the protein in the actin changes shape and allows the cross-bridges to form.

You have more than 600 muscles in your body, arranged in groups or pairs. Some of them work together (synergistically) to produce movement. Others work against each other (antagonistically) to produce movement.

When muscles contract, they transmit the force of the contraction to the bones to which they're attached. Taken together, your skeleton and muscles work like a system of levers in which rigid rods (your bones) move around at fixed points (your joints). Most of your muscle attachments are close to your joints, so this means that your muscle needs to move only a small amount to produce a large movement of a body part.

Your skeletal-muscular system is one of the most fascinating systems in your body—fascinating because you can physically see it in action every day (every second, if you like). Think about all the coordinated activity that's involved just in picking up a glass and getting it to your mouth, and even the muscles in your lips and mouth that enable you to swallow. Even crawling into bed at the end of a long, strenuous day requires bones and muscles. In the next chapter, we'll be discussing your nervous system, the system in charge of all the commands in your body.

Weird Science

When you die, ATP production stops. Without ATP, the cross-bridges between the myosin and actin never detach. All the cross-bridges remain locked in place, and the skeletal muscles of the body become rigid, which is known as rigor mortis.

Think About It!

If you suffer from "twitchy legs" or uncontrolled muscle spasms in your limbs, what can you do to get control over those contractions? Consume more calcium!

The Least You Need to Know

➤ The integumentary system (your skin) of vertebrates helps protect the body from dehydration, bacterial attack, ultraviolet radiation, and abrasion. Your skin helps control your internal temperature, serves as a blood reservoir, and detects environmental stimuli.

➤ Bones provide the structure of vertebrate skeletons. Your bones protect and support other organs, interact with your muscles for movement, store minerals, and help in the formation of red blood cells.

➤ All muscle tissue is excitable, meaning that it responds to stimulation, can contract, and is elastic. Your body has three types of muscle: smooth, cardiac, and skeletal. Only skeletal muscle interacts with your bones to bring about movement in the environment.

➤ Skeletal muscle cells contain threadlike myofibrils, which contain actin and myosin filaments. These filaments are organized into orderly sarcomeres, the basic units of muscle contraction.

➤ Your skeleton and muscles work like a system of levers, in which rigid rods (your bones) move around fixed and flexible points (your joints).

Chapter 12

The Nervous and Sensory Systems

In This Chapter

➤ You're nothing without organization.

➤ The control center: the brain

➤ The pathways: neurons

➤ The messengers: nerve impulses

➤ The five senses, and more

When it comes to your body's systems, it's hard to rank them in order of importance. Every system has an integral role to play in keeping your body up and running. But of all your systems, your nervous system is definitely the commander-in-chief. Ultimately, it may not be more important, but it plays a vital role in the way all the other systems of the body function. Your nervous system is responsible for sensing and interpreting what's happening in the environment. Then, in response, it issues commands to all of your body's systems. Cougar in your path? Thank your nervous system for getting you the heck out of there.

In animals, the nervous and sensory systems operate together, providing the framework for an organism's survival. You'll see what we mean in this chapter.

Always, Always, Always: Organization!

As a group, animals have highly organized and complex nervous systems. Sponges are about the most simplistic members of the animal kingdom, and even they have a nervous system, albeit a pretty simple one. Prick a sponge with a needle, and it will respond by slowly contracting a few millimeters away from the source of stimulation. It's not a very impressive or lightning fast response, but it is a nervous response.

As a general rule, the more complex the organism, the more complex the nervous system, which is why humans have the most complex nervous system of all known animals.

Invertebrate Organization

Invertebrates with radial symmetry, such as hydras, sea anemones, and other cnidarians (radial invertebrates, such as jellyfish), exhibit the simplest nervous systems. If you recall, their body parts are arranged around a central axis, like the spokes on a bicycle wheel. These type of invertebrates have a nervous system that radiates out through the "spokes," allowing the animal to sense danger (or food) coming from every direction. The type of nervous system that radial invertebrates have is called a *nerve net*.

Suppose that a jellyfish wants to eat. What nervous system pathways in its nerve net are involved? In the pathway concerned with feeding behavior, nerve cells extend from the sensory receptors in the jellyfish's tentacles to the contractile cells around its mouth, which allows the jellyfish to sense the presence of its food, get the food to its mouth, and then contract around the food to "swallow" it. Reflex pathways also permit the jellyfish to swim slowly and keep its body upright.

Bio Buzz

A **nerve net** is a type of nervous system found in cnidarians. It is comprised of sensory cells, nerve cells, and contractile cells. All three cell types interact to form reflex pathways.

Bilateral and Cephalized Invertebrate Organization

Flatworms are the simplest of all bilateral and cephalized invertebrates, and even they have an organized nervous system. Bilateral nervous systems probably evolved from nerve net systems because both systems show some similarities during their larval stage of development. But rather than being a net structure, bilateral nervous systems are characterized by two equal parts and a "front end."

The front end of a forward-crawling animal (aquatic or terrestrial) is the first body part to encounter food and external stimuli. Consequently, the front end of cephalized animals is the center of the nervous system; it forms the beginning of a brain, if you will. All bilateral, cephalized invertebrates have a brain, a nerve cord, and ganglia (small clusterings of neuron cell bodies).

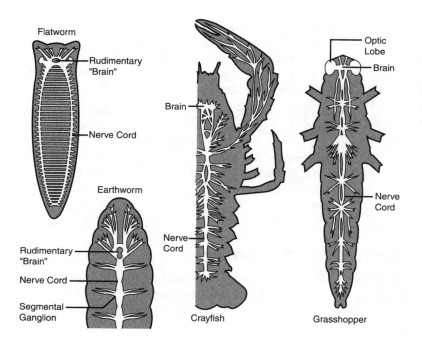

Flatworms, earthworms, crayfish, and grasshoppers all exhibit highly organized, bilateral, cephalized nervous systems composed of a brain, a nerve cord, and ganglia/neurons.

Vertebrate Organization

All vertebrates exhibit bilateral cephalization, so it probably comes as no surprise that the same patterns found in the nervous systems of bilateral cephalized invertebrates exist in the nervous systems of vertebrates, specifically humans. Vertebrate nervous systems show much more organization, bilateral symmetry, and cephalization than do invertebrate systems. Like invertebrates, the first vertebrates had a nerve cord. But from there, things started evolving.

In vertebrates, the nerve cord was the forerunner of the brain and the spinal cord. Millions of years ago, in the changing world of vertebrates, any species that was better equipped to sense and respond to its environment had a competitive edge in terms of survival. In vertebrates, the backbone evolved millions of years ago and was key to the development of speedy, jawed predators. As the senses of hearing, sight, smell, and balance further evolved, the brain itself had to evolve in order to be able to assimilate all the sensory information it received. Over time, these thickening brain tissues developed into three distinct regions: the forebrain, the midbrain, and the hindbrain (once again, originality strikes in the scientific names department).

Think About It!

For an interesting and thought-provoking read on the evolution of human intelligence, pick up a copy of Carl Sagan's book, *The Dragons of Eden* (Ballantine, 1977). It's a fascinating look at our own "inner space."

Vertebrates still retain some ties to their ancient ancestors. Today, the *nerve cord* still shows up in all vertebrate embryos but is now called the neural tube. During embryo development, the neural tube expands and modifies into the brain and spinal cord.

For descriptive purposes, vertebrate nervous systems are organized into two main parts: the central nervous system (which includes the spinal cord and brain) and the peripheral nervous system (which includes all the nerves that carry signals to and from the spinal cord and the brain).

The Command Center: The Brain

All vertebrates and most invertebrates have a brain. It may not be your notion of a brain, but scientifically speaking, it's a brain. There are all sorts of brains, but in keeping with the narcissistic tone of this book, we'll confine our discussion to the most interesting thing on the planet: you.

Brains are the General Patton of the body. They gather intelligence, peruse all the data, make decisions, and marshal the troops into action.

Let's look at an example of your brain in action. Imagine that it is a beautiful late-summer day in Kansas. The wild flowers are blooming profusely, and the bees are humming. Nothing mars the day, except perhaps some large clouds developing on the horizon. You're checking out the wildflowers in your guidebook and enjoying the day in general, so you fail to notice the ominously darkening sky. When you finally do look up, it's because of a sound like a rumbling freight train. But there are no railroad tracks around for miles and miles. What's making that noise?

Then you see it, but at first it doesn't register. It's a tornado! And it's heading straight toward you!

Your senses immediately click into high gear. You know that you can't stay where you are and survive. You remember that a low spot is better than an open area, but where's a low spot? No matter—your brain flashes out commands, and you start to run as fast as you can. The commands are loud and clear: Get out of here!

With your heart pumping and adrenaline surging, you thunder along the path, desperately looking for a low spot. Just ahead is a small creek. With a mighty leap, you land on the cobblestones and scramble under the muddy over-hanging bank. It takes a few minutes for you to realize that the tornado has passed over and that you escaped unharmed. You unclench your fingers from the mud and crawl back up the bank. In the distance, you see the funnel cloud disappearing over a hill. Your heart is no longer pounding, and your blood is no longer rushing. In fact, you feel a bit shaky and plunk down on the grass.

What got you through this tornado trauma? Your nervous system. Every perception, thought, and memory came from your nervous system. Every physical action that your skeleton and muscles performed was directed and driven by your nervous system. Were all these communication lines that directed your actions and thoughts silent before the tornado appeared in your vision?

Hardly. The vast network of your nervous system continually jabbers away, frequently just to itself. Constant communication between the commander in chief (your brain) and the troops (the neuron network of your sensory and peripheral nervous systems) is what keeps you alive.

The human brain resembles a walnut, in that it is crinkled and folded in appearance. (Other than that, it has no relationship to a walnut, although some people might beg to differ.) The folds of your brain suggest that its evolutionary physical development outpaced the development of your skull bones. To accommodate all this growth in a confined space, the human brain began folding in upon itself in order to fit inside your skull.

Your brain starts out as the continuation of the anterior end of your spinal cord. Like your spinal cord, it's protected and enclosed by bones and tissues. A deep fissure divides your brain into two parts: the left and right cerebral hemispheres. Each of the hemispheres has its own set of communication lines among its different regions.

The functioning of the cerebral hemispheres has been the subject of numerous experiments. Each hemisphere can function by itself, but between them, there is a division of labor. The main regions responsible for spoken language skills are found in the left hemisphere. The main regions responsible for nonverbal skills, such as music, mathematics, and abstract reasoning, are located in the right hemisphere. The corpus callosum, a band of about 200 million neurons, allows the two hemispheres to communicate with each other.

In addition to your two hemispheres, your brain has three other regions: the hindbrain, midbrain, and forebrain. The hindbrain consists of the cerebellum, pons, and medulla oblongata. The hindbrain is located in the lower back portion of your skull and is essentially the continuation of your spinal cord.

The medulla oblongata controls your breathing, blood circulation, and other vital bodily functions. This is also where the complex motor responses, such as coughing, are located. The centers in the medulla oblongata influence the other parts of your brain that help you sleep or wake up.

Try It Yourself

Your senses and brain intertwine with other systems in your body. For a demonstration of this, go to a restaurant when you're moderately hungry. See what happens when you smell the food. Your sense will pick up the odor (the information), your brain will relay it (instantaneously), and your stomach will probably tell you that—all of a sudden—it's hungry, very hungry.

The cerebellum helps maintain your posture and refine your limb movements. It keeps the brain informed about what's going on with your limbs, and it coordinates signals from your eyes, skin, muscles, and elsewhere.

The pons is the traffic control center for all the nerve networks passing between your brain centers.

As its name suggests, the midbrain is the middle portion of your brain, located above the hindbrain. In animals such as reptiles, fishes, amphibians, and birds, the midbrain integrates signals from the eyes and ears, which is a vitally important function. In humans, the midbrain performs much the same service: It coordinates your reflex responses to visual and auditory stimuli. Taken together, the midbrain, medulla oblongata, and pons comprise what is known as the brain's "stem."

The forebrain contains the most highly developed regions of your brain. One center, the cerebrum, coordinates odors and your responses to those odors.

Another center, the hypothalamus, monitors your internal organs and behaviors related to them, such as thirst and hunger. The *cerebral cortex*, yet another center, encodes information.

The spinal cord is the expressway for signals between your brain and the peripheral nervous system. All the signals from your sensory organs and neurons throughout your body travel along the spinal cord before ultimately reaching the brain. Your spinal cord threads through the stacked bones of your vertebral column, or backbone. The bones, and the ligaments protecting the bones, enclose and protect the spinal cord.

Bio Buzz

The **cerebral cortex** is actually a thin surface layer of cells covering your cerebral hemispheres. Some regions of your cerebral cortex receive sensory input, while others integrate the information and direct the appropriate motor response.

Both your brain and your spinal cord are incredibly fragile—imagine cradling Jell-O in your hands and trying not to damage it. That's roughly the equivalent of how fragile your brain is. Together, the brain and spinal cord comprise your central nervous system.

The Pathways: Neurons and the Peripheral Nervous System

The brain might be controlling everything, but how does it get the information that enables it to make decisions? The peripheral nervous system delivers most of the messages your brain receives.

In humans, the peripheral nervous system consists of 31 pairs of spinal nerves that connect directly to the spinal cord. It also has 12 pairs of cranial nerves connecting directly to the brain. Some nerves in your peripheral system carry only sensory information, such as the optic nerves in your eyes, which carry information from your eyes to your brain. Other nerves have a combination of both sensory and motor axons. For instance, vagus nerves are found in your abdomen. They have sensory axons leading into the brain, as well as motor axons going to the heart, lungs, and gut.

Your peripheral nervous system has two subdivisions: somatic and autonomic. Your somatic peripheral nervous system deals with things you can control, such as the movement of your limbs, trunk, and head. Sensory axons found in this system pick up signals from receptors in your skin, skeletal muscles, and tendons. In turn, motor axons carry signals out to your skeletal muscles.

The autonomic peripheral nervous system, on the other hand, deals with the functions of your internal organs and structures, which you can't control. Sensory and motor axons of this system carry signals to and from your cardiac muscle, smooth muscle, and different glands in different regions of your body.

Autonomic nerves are further divided into two more subdivisions of nerves: parasympathetic and sympathetic. Parasympathetic nerves tend to send signals that slow down your body and divert energy to your body's basic housekeeping tasks, such as digestion. Sympathetic nerves tend to increase the level of bodily activity, preparing you to fight, flee, or play.

Both the parasympathetic and sympathetic nerve systems work together: If one system decreases suddenly, the other may rise up with equal rapidity. For instance, the stimulus of a tornado causes your sympathetic system to spring into action. Yet when that stimulus disappears and the tornado moves away, your sympathetic system drops precipitously, and your parasympathetic system rises dramatically, causing you to feel weak and rubbery-legged, or even to faint. Essentially, your nervous system is out of sync during this period.

Try It Yourself

You can perform this simple experiment with your sense of taste in your own kitchen. All you need is some things with distinctly different tastes: sugar, lemon juice, salt, vinegar, and others. (The idea is to taste, not eat.)

One at a time, take a little bit of each item you've selected to taste. Swish it around your mouth and wait for a reaction. You'll obviously taste something immediately, but after the first sensation, pay closer attention to where the most taste is coming from. Is sour at the front, back, or sides of your tongue? Is sweet at the back or front?

Your taste buds aren't randomly scattered over your tongue. Different areas of your tongue will pick up different tastes better than other places.

Both kinds of autonomic nerves (parasympathetic and sympathetic) are continually chattering to each other, carrying signals to and from your central nervous system. How do all these nerves "talk" to each other and relay information? That's where the nerve impulse comes in.

The Messengers: Nerve Impulses

In all nervous systems, vertebrate or invertebrate, the basic unit of communication is the *neuron*, or nerve cell. Just as no man is an island, neurons never stand alone. They work collectively to sense changes in your body and the environment, to integrate and coordinate all the incoming signals from your sensory receptors, and to activate the appropriate body parts needed to carry out a response.

There are three main types of nerve cells: sensory neurons, interneurons, and motor neurons. To accomplish tasks, all three types of neurons exchange signals with each other.

➤ **Sensory neurons** are equipped with receptors or receptor regions that detect specific stimuli such as light, heat, and cold. Sensory neurons relay signals *to* the brain and spinal cord; they operate in only one direction.

➤ **Interneurons** are found in the brain and the spinal cord. They integrate all the information coming in from the sensory neurons and influence other neurons in turn.

➤ **Motor neurons** relay information *away from* the interneurons out to your body's muscle cells and gland cells. Your muscles and glands carry out the necessary responses. Like sensory neurons, motor neurons work in only one direction. Between the two of them, they form a highway that runs both directions.

The three classes of neurons (sensory, interneuron, and motor) comprise about half the volume of your nervous system. Neuroglial cells are also present in great numbers. *Neuroglial cells* are specialized cells that perform many supportive functions. They physically support and protect the neurons, helping them to carry out their tasks. Some types wrap around and protect neurons, to the extent that they actually segregate groups of them. In the brain, neuroglial cells provide structure for brain tissues, much like connective tissues do in the rest of the body. Neuroglial cells also affect how fast a signal travels along the neuron pathway.

Bio Buzz

A **neuron** is the basic unit of communication within your nervous system. Neurons, or nerve cells, collectively sense changes, integrate the information, and initiate your body's response.

Functional Parts of Neurons

Because neurons are cells, each of them has a cell body, complete with a nucleus and the capacity for protein synthesis. But when it comes to physical appearance and structure, there is no such thing as a "typical neuron." Somewhat like snowflakes and fingerprints, each type of neuron is unique. The most commonly described neurons are motor neurons, so for discussion purposes, we'll talk about them.

Bio Buzz

Neuroglial cells provide structural and metabolic support for neurons in vertebrates. Collectively, neuroglial cells represent about half the volume of vertebrate nervous systems.

In addition to a cell body, motor neurons have short, slender, threadlike extensions branching out of their cell body. These extensions are called dendrites. Motor neurons also have one long, slender, cylindrical extension, which looks a bit like a tail, called an axon. In motor neurons, the ends of axons are finely branched and connect to muscle cells. Dendrites and axons are also found on many sensory neurons and interneurons.

In general, dendrites are the "input zone" of a neuron, the place where signals are received, and axons are the "output zone" of the neuron, the place where signals are relayed on to the next cell.

Send Those Messages Down The Line!

The messages that neurons receive and pass on are actually signals. Some signals are electrical, while others are chemical. But no matter the type of signal, it transmits information to the neuron.

A resting neuron, one that's not doing anything in particular, exhibits a steady and constant difference in the electric charge that's across its plasma membrane. In short, there's a voltage difference between the interior of the plasma membrane and the exterior. The cellular fluid inside the cell is more negatively charged than the fluid on the exterior of the plasma membrane. This steady difference in voltage is called the *resting membrane potential*; it's the cell's potential for activity.

To follow the progression of a message along a neuron chain, let's look at the illustration of a clumsy person (you) and a sleeping cat. Suppose that your cat is snoozing quietly in some out-of-the-way place, like the middle of your kitchen floor. Now let's suppose that you just happen to stumble over this cat, stomping its tail in the process. What's going to happen?

In stepping on your cat's tail, you start a tidal wave of signals coursing through your cat's nervous system. And although your cat is a cat, and you are a human, you're both vertebrates, so your neuron processes are virtually the same. If your cat wants to turn around and stomp on your toe, the same set of signals will go off in your nervous system.

As the signals travel from the sensory neurons, into the interneurons, and on through the lines, so to speak, your cat's motor neurons suffer a disturbance.

The same thing goes for you. The information, or signal, that travels along your neurons is just a transient change in the voltage charge of the cell's plasma membrane. Weak disturbances might trigger a tiny change across the plasma membrane, one so slight that your body barely acknowledges it.

Strong disturbances, though, are another matter. In the case of your cat's tail, the disturbance is strong enough to trigger a nerve impulse (sometimes called an action potential). For just a fraction of a second, the interior of the neuron cell becomes positive, with respect to the exterior of the cell. This change in the cell's resting membrane potential travels out along the length of the axon.

When the nerve impulses finally reach the axon endings of the motor neurons in your cat's tail, they trigger a chemical signal to the adjacent muscle cells. In response to this signal, your cat's muscles contract, and your cat most likely shoots straight up in the air (among other things).

Bio Buzz

The **resting membrane potential** is the steady voltage difference across the plasma membrane of a resting neuron. The fluid interior will be more negatively charged than the fluid exterior of the plasma membrane.

In all neurons, stimulation of the dendrites (the input zone) produces a localized signal that doesn't spread very far. In your cat's case, you initially disturbed the sensory neurons with dendrites embedded in the connective tissue of your cat's skin. At each receptor site, mechanical pressure (your foot) caused the change in voltage across small patches of the plasma membrane. This voltage change produced a graded, local signal.

Your nervous system grades signals based upon their magnitude. Signals are graded large or small, depending upon the duration and intensity of the stimulus. Your foot has probably created a signal that your cat's nervous system grades as large. The greater the intensity and duration, the greater the disturbance to the sensory neurons.

If the signal is weak or small, it doesn't go very far. But the stronger the signal, the more likely it will spread to an adjacent trigger zone, which is the membrane site where nerve impulses are created.

When you stepped on your cat's tail, you triggered hundreds of nerve impulses, all in the single second that your foot came down. When signals cross the threshold level, voltage-sensitive gates for sodium ions opened in the neuron cell's plasma membrane. The inward flow of these positively charged sodium ions makes the interior of the cell less negatively charged. This causes even more gates to open and more sodium ions to enter, until the charge reverses.

Once the threshold is reached, this cycle of more sodium gates opening no longer depends upon the strength of the stimulus; it proceeds automatically. That's why in the case of a large stimulus, your neurons spike at the same time, in an "all-or-nothing" event.

This initial voltage reversal travels along the neuron pathway, with one axon leading into another dendrite, and sodium gates opening and producing more signals—eventually (a split second later) your cat's muscles respond.

Weird Science

The nerve pathways of vertebrates, and their operation, are the most complex of all animals. Besides being a vast, nearly impenetrable network within your body, your nerve cells are different from other cells because they can't repair damage to themselves. That is why people who suffer spinal cord or brain damage sometimes are confined to wheelchairs for life. Their muscles are fine, but no signals get through because the pathway is damaged.

The Five Senses, and More

If you think of your nervous system as a mansion full of complex rooms and corridors, your sensory system is the front door. Your sensory system provides your nervous system with the raw data that the brain needs to give your body the information it needs to survive. Sponges are about the only animals without a sensory system that still manage to survive. Every other animal needs a sensory system. Imagine a gazelle trying to eat and avoid predators on the African plains. Or, for that matter, imagine yourself negotiating rush hour traffic. Animals need complex sensory systems just to handle the huge volume of information that comes in every second. Without an elaborate sensory system, most animals couldn't survive.

Every sensory system is composed of three parts:

1. Sensory receptors

2. Nerve pathways

3. Brains, or brain regions, where the information is processed

Bio Buzz

A **sensation** is a conscious awareness of a stimulus.

Sensory receptors are the finely branched endings of sensory nerves. Their purpose is to detect specific conditions, or stimuli. The stimulus could be heat, light, pressure, or just about anything capable of eliciting a response from the receptor. Once a sensory receptor detects the stimulus, the energy from the stimulus is converted into a nerve impulse; in that form, the information travels along the nerve pathways until it arrives at the brain. The brain translates the information into a specific *sensation*.

Sensation is not the same as perception. Just because you're aware of a sensation doesn't mean that you necessarily understand it.

Types of Sensory Receptors

Sensory receptors are specialized, in that many of them pick up only a certain type of stimulus. Most of them are individually positioned in your skin and body tissues. Some are part of other organs, like your eyes, and serve to amplify or concentrate whatever the stimulus energy is.

There are five types of sensory receptors, and scientists group them according to the type of stimuli they receive.

1. **Chemoreceptors** detect the chemical energy of specific substances that are dissolved in the fluid surrounding the receptor.

2. **Mechanoreceptors** detect the mechanical energy that's associated with changes in pressure, acceleration, or changes in body position.

3. **Photoreceptors** detect light, both visible and ultraviolet.

4. **Thermoreceptors** detect heat.

5. **Nociceptors** are pain receptors. They detect tissue damage.

Not all animals have the same kinds or number of sensory receptors. Different animals sample the environment in different ways. For instance, you can't see ultraviolet light, but a bee can. You don't have the mechanoreceptors necessary to detect ultrasound waves, but bats and whales do. And you don't have thermoreceptors that let you detect the heat of a warm-blooded animal in the dark, but a python does.

Each nerve pathway starts with sensory receptors that are sensitive to the same type of stimulus, whether pressure, heat, light, or some other stimulus. The nerve pathways from all these different receptors lead to different parts of the cerebral cortex, which is the outermost layer of the brain. The areas of your brain that receive all these incoming messages are laid out like a map, and different areas receive only certain messages.

Nerve impulses are not like a wailing ambulance siren; they don't vary in amplitude. When they arrive in the brain, what message are they delivering?

First, the brains of animals can interpret nerve impulses only in a specific manner. An impulse from a specific sensory receptor going to a specific region of the brain will always be interpreted in the same manner. That's why you "see stars" if you poke your eye in the dark. The region of your brain that receives stimuli from your eyes always interprets that stimuli as light.

Second, if a sensory receptor is receiving strong stimulation, the nerve impulses will fire off more rapidly. The brain responds to this frequency.

Third, if a receptor is receiving strong stimulation from one area, it will recruit more receptors, essentially covering a larger area. A light touch on your head will activate some receptors, but being whacked on the head with a beer bottle will activate a lot more receptors.

Sometimes the frequency of the nerve impulses will decrease or stop altogether, even when the stimulation is being maintained. When you first put on your clothing, your body registers the stimulation, but then it stops and you don't pay any attention to your clothing (usually). The clothing is still there, and the stimulation is still there, but the nerve impulses have gone home. This occurrence is called adaptation.

Somatic Sensations and the Special Senses

Based on your sensory receptors and where they're located, scientists group sensations into two rough groups: somatic and special.

Somatic sensations include pain, cold, heat, pressure, and touch. They also include such things as the awareness of your body's position in space and the positioning of your limbs. These sensations come from sensory receptors that are embedded in your skeletal muscles, your internal organs, and your skin and other tissues.

The special senses include taste, smell, hearing (which in itself includes echolocation and balance), and vision. Animals show great diversity when it comes to the special senses. For instance, all animals can hear, but many of them don't possess "ears." In grasshoppers, the sensory receptors that pick up sound waves, or hear, are located on their hind legs. In crickets, they're located on the front legs.

No matter what the variety, all animals have nervous and sensory systems. Without both of them, working together, animals couldn't survive.

We've covered a lot of ground in this chapter, but how could we not? Any nervous system is complex, and yours is especially intricate. From that first sense to the marshaling of a response from your brain, to nerve impulses moving down the line, your nervous system is an amazing evolutionary feat. In the next chapter, we'll be talking about something everyone likes: food, and how your body uses it.

The Least You Need to Know

➤ The nervous system of animals provides a specialized way of detecting changes (stimuli) and responding to them.

➤ The simplest nervous systems are nerve nets. They are found in jellyfish, hydra, sea anemones, and other such animals.

➤ The brain and spinal cord represent the central nervous system. Pairs of spinal nerves carry information to the brain, where response decisions are made.

➤ The peripheral nervous system spreads out through the body, gathering information to be funneled to the spinal cord and relayed to the brain.

➤ The sensory system is the door into the nervous system. The sensory system is the front line for gathering stimulus information and commencing the information pathway.

Chapter 13

The Digestive System

In This Chapter

➤ Everyone needs organization

➤ Down the hatch: digestion

➤ Use it: absorption

➤ You are what you eat: nutrition

"You are what you eat." How many times have you heard that particular platitude? In the age of super fitness and spandex bodies, it sounds like just one more mantra to chant while pushing a fudge brownie out of reach.

But there's a great deal of truth to the phrase. An animal's digestive system connects it directly with its environment and keeps it alive. Every living organism needs nourishment. If you want to keep being you, you need to eat.

Within the animal kingdom, there's great diversity among digestive systems, but all of them are highly organized and possess common traits. We'll discuss that organization in this chapter.

Always, Always, Always: Organization!

In general, a digestive system involves some type of tube or body cavity in which food is first reduced to small particles, and then is further reduced to small molecules. A layer of cells lines the tube or cavity, and nutrients enter the internal environment of the animal's body by crossing over this cell lining.

Incomplete Digestive Systems

A lot of invertebrates have extremely simple, *incomplete digestive systems*. The digestive systems of corals, sea anemones, flatworms and other invertebrates, for instance, consist of only a saclike gut with one opening.

Bio buzz

An **incomplete digestive system** is a digestive system in which the gut has only one opening. Food enters and waste products leave through the same opening.

In most animals with an incomplete digestive system, a muscular pharynx opens into a digestive cavity with a branching network. This cavity and its branches do duty for both digestive and circulatory functions. With an incomplete digestive system, food is being consumed and digested even as waste products are being sent back out through the pharynx.

Scientists think that this two-way traffic situation must have worked against the development of a more sophisticated system of digestion during the evolutionary process, preventing specialized regions for food transport, processing, and storage from developing.

Even the most "incomplete" incomplete digestive systems, though, have the basics: a body cavity lined with cells for transporting nutrients.

Complete Digestive Systems

Chordates and vertebrates such as humans, moose, hummingbirds, and crocodiles have what's called a "complete digestive system." Annelids, mollusks, arthropods, and echinoderms also have complete digestive systems.

A complete digestive system takes in food through an opening at one end and disposes of wastes through an opening at the other end. In between, the food travels through a tube that is divided into specialized regions for food transport, processing, and storage. Between the two openings, food generally moves in one direction, working its way through the lumen, or space inside the tubular structure.

Complete digestive systems have four components, or traits:

➤ **Motility**—Complete digestive systems have muscular gut walls, which move the food through and aid in the breakdown, mixing, and nutrient dispersal. Motility also helps to eliminate wastes and undigested food.

➤ **Secretion**—The cells that line the lumen of complete digestive systems are constantly secreting enzymes, fluids, and other substances that aid in the processing, absorption, and other functions of the digestive tract.

➤ **Digestion**—All complete digestive systems break down small particles of food into even smaller molecules that can then be absorbed by the organism.

➤ **Absorption**—Complete digestive systems are characterized by their ability to transport nutrients, fluids, and ions across the tube wall and into the blood or lymph systems, where they are distributed or absorbed by the rest of the body.

A "complete digestive system," then, is really a term for a general grouping. Within that grouping, a wide variety of different types of digestive systems exist. The digestive system of birds, for instance, is modified into a crop, or food storage organ. Another part of the system is modified into a gizzard, which is a muscular organ that grinds food into smaller pieces that are then further digested. You won't find a crop or gizzard anywhere in your own digestive system, yet both you and birds possess complete digestive systems.

Weird Science

If your digestive tube were stretched out from end to end, it would stretch somewhere between 6.5 and 9 meters (approximately 21 to 30 feet).

Your digestive system is vastly different from any of these other mammals, and nothing like that of an echinoderm. Humans have discontinuous feeding habits, and they also ingest a wide variety of food. Ruminants eat only vegetable matter, and predators tend to stick with meat. But in any given day, a Tibetan monk might eat only rice, an Eskimo might eat only raw seal blubber, and an American might eat sausage pizza, artichokes, ice cream, pineapple, and oatmeal. Like magic, the human digestive system will handle all of it, converting it into useable energy and tissues of its own. Because of our feeding habits, the human digestive tube has many specialized regions.

The specialized regions of the human digestive system start with an oral cavity, or mouth. From there, food passes through the pharynx, esophagus, and the gut (gastrointestinal tract). The gastrointestinal tract is further subdivided into the stomach, small intestine, large intestine, rectum, and anus. From salivary glands in your mouth, all the way through the system, enzymes are continually secreted, which aids in digestion and absorption.

All digestive systems are interesting, but for the rest of this chapter, we'll concentrate on the human digestive system.

Down the Hatch: Digestion

Food enters your body through your mouth, or oral cavity. In this particular case, your mouth is more than just an intake valve; it's the start of the digestion process. The moment food enters your mouth, it gets assaulted by your teeth and attacked by starch-degrading enzymes.

Saliva (and your teeth) starts the digestion process. Saliva is a fluid secreted by your salivary glands, which are located beneath your tongue and in the back of your oral cavity.

Saliva contains salivary amylase (a starch-degrading enzyme), bicarbonate, and mucins (modified proteins). The buffering action of the bicarbonate keeps the interior of your mouth at a pH of somewhere between 6.5 and 7.5, even when you're eating acidic foods, such as tomatoes.

The mucins bind the bits of food into a softened ball, called a bolus, and muscle contractions of your tongue force the bolus into your pharynx. The pharynx is a muscular tube connected to the esophagus, which leads directly into the stomach. You swallow when you voluntarily push a bolus into your pharynx. This stimulates sensory receptors in your pharynx, which in turn trigger involuntary muscle contractions, which starts the bolus moving downward.

Motility and Muscles

Your pharynx and esophagus don't play direct roles in digestion because neither of them break down food or aid in the dispersal of nutrients. But they're both instrumental: The muscle contractions of their walls are what propels food into your stomach.

Weird Science

Humans and other mammals are the only animals that chew their food. To accomplish this task, adult humans generally have 32 teeth. Chisel-shaped incisors bite, pointed cuspids rip, and flat-topped molars grind. Each tooth has a hardened enamel coat of calcium deposits and is designed to withstand years of mechanical stress and chemical wear and tear.

Your digestive system is more than just a slimy tube; it's composed of muscles from top to bottom. Remember the three types of muscles in your body (cardiac, smooth, and skeletal)? Smooth muscles are found in your digestive tract. You can move them voluntarily, as you can your skeletal muscles, but smooth muscle is muscle, nonetheless.

The structure of your gut wall is a series of layers. *Mucosa* is the innermost layer that faces the lumen. Next comes the submucosa, a layer of connective tissue interwoven with blood and lymph vessels, and local neuron networks. The third layer from the lumen is the smooth muscle. Smooth muscle usually forms in two layers, each running in a different direction. One layer of smooth muscle runs in a circular direction, while the other layer runs longitudinally, or up and down.

Sphincters occur at the beginning and end of your stomach, as well as in other specialized regions of your system. Sphincters help control the forward movement of food and prevent it from flowing back into the digestive system.

The muscles of your digestive system produce wavelike contractions that mix food and move it through the whole system. During peristalsis, the ring muscles contract behind the food, while relaxing in front of it. This motion pushes food forward, moving it along. As the food moves forward, it distends the tube walls, and peristaltic motion keeps pushing it along. During segmentation, rings of smooth muscle repeatedly contract and relax. This type of movement, which is similar to the back-and-forth motion of a washing machine, mixes the food and pushes it against the absorptive surface of your gut walls.

Fill That Stomach!

Your stomach is a muscular, flexible sac, directly connected to your esophagus. When you swallow, food goes into your pharynx, then your esophagus, and then right into your stomach.

Your stomach serves three functions in your digestive system:

1. It stores and mixes food.
2. It secretes enzymes and acids, which help dissolve food.
3. It controls the movement of food into your small intestine (where absorption will take place).

The storage and mixing of food is a pretty clear-cut function: You swallow, the food reaches your stomach, and mixing commences. Secretion is a little more complex. Every day, the cells in your stomach lining secrete approximately 2 liters of fluids, including hydrochloric acid, mucus, and pepsinogens. Together, these substances make up your gastric fluid.

The hydrochloric acid separates into H+ and Cl-. This increase in acidity helps dissolve food. It also manages to kill most bacteria that have hitched a ride into your digestive system. The dissolving food and acids form a mixture referred to as chyme.

Your stomach is typically the most acidic part of your body. How could it not be, with the constant secretion of hydrochloric acid? Stomach secretion starts with your brain (it's those good ol' sensory receptors again). Your brain responds to the smell, sight, and taste of food. It even responds to hungry thoughts about food. In response, the brain fires off signals to the cells lining your stomach. This starts the process of secretion.

Bio Buzz

The **mucosa** is the innermost layer of your gut. It faces the lumen and is composed of epithelium, with an underlying layer of connective tissue. A **sphincter** is a circular ring of muscle that occurs in the wall of human digestive systems.

Weird Science

Your pharynx also connects with your trachea, which leads directly to the lungs. When you swallow, you open a sphincter at the start of the esophagus. A small, flap-like valve, the epiglottis, then closes off your respiratory tract, which is why you're not continually choking on your food every time you eat.

Think About It!

Watch your caffeine intake! Your stomach varies its secretions based on actual substances. For instance, your stomach's sensory cells really step up secretions when they encounter the caffeine in coffee, tea, chocolate, and soft drinks.

Weird Science

Sometimes your stomach's mechanisms go awry. The surface of your stomach breaks down, and H+ diffuses into the lining and triggers the release of histamine. Histamine stimulates the production of even *more* hydrochloric acid, and a positive feedback loop is established. The lining becomes further damaged and starts bleeding. This condition is called a peptic ulcer.

When food actually arrives in your stomach, the cells respond to the stretching of the stomach sac and step up the secretions. Protein is harder to digest than other substances, such as carbohydrates. Stomach acid is the first major step in breaking down proteins so that your body can use them. The acid breaks down the structure of protein and exposes the peptide bonds. The acid also converts pepsinogens to active forms that are able to break down the proteins even further. As the protein becomes fragmented, the protein fragments themselves directly stimulate the production of gastrin, which in turn stimulates the secretion of more hydrochloric acid. So, the more protein you eat, the more gastrin and hydrochloric acid forms in your stomach.

With all those caustic substances continually swishing about in your stomach, what keeps your stomach intact? Why doesn't the lining just disintegrate under the acidic assault?

Your stomach (and brain) have control mechanisms to ensure that there is always enough mucus and buffering molecules (particularly bicarbonate) present in your stomach to protect the lining from the destructive effects of hydrochloric acid.

The peristaltic waves of your stomach muscles mix the chyme and build up force, particularly as they approach the sphincter at the end of your stomach, leading into your small intestine. A strong contraction will close off the sphincter, forcing the chyme back into the stomach. But little by little, the chyme begins making its way into the small intestine.

The volume and the composition of the chyme in your stomach determines how fast your stomach will empty into your small intestine. Large meals distend your stomach and send out signals to increase the contractions. Likewise, your small intestine can sense increases in acidity, fat, and other things, and it sends out signals to slow down the emptying of your stomach. By working together, your stomach empties only as fast as your small intestine can process the chyme.

Use It: Absorption!

Your small intestine is where digestion is completed and nutrients are absorbed and shipped out to your body for use. The small intestine has three regions: the duodenum, the jejunum, and the ileum. The pancreas and liver also contribute to the functioning of the small intestine. A common duct empties secretions from both these organs into your small intestine.

Your pancreas aids in digestion by secreting enzymes that help to digest fats, proteins, carbohydrates, and nucleic acids. Just like the pepsin your stomach secretes, the pancreatic enzymes trypsin and chymotrypsin help your small intestine break down proteins. The pancreas also secretes bicarbonate, which buffers and helps to neutralize the hydrochloric acid arriving from your stomach. The pancreas also secretes two other hormones, insulin and glucagon, which don't help with digestion but are important in a nutritional sense.

Your liver plays a key role in digestion because it secretes bile. Bile salts help with the breakdown and absorption of fat. Most of the fats found in the typical American diet are triglycerides, which clump together in large globules. The small intestine breaks down these globules mechanically through its muscle action. Bile salts help keep these smaller droplets from reclumping into a large globule. This process is called emulsification. The emulsifying effect of bile salts on fat globules lets the fat-degrading enzymes in your small intestine get better access to the triglycerides, which enhances fat digestion. Between meals, when your small intestine doesn't need *bile*, your gallbladder stores and concentrates the bile.

By the time a meal has made its way through half of your small intestine, proteins, lipids, and carbohydrates have all been broken down into small molecules, thanks to the mechanical action of muscles and the chemical action of enzymes. This breakdown produces other products, such as glucose, amino acids, fatty acids, monoglycerides, and monosaccharides. All of these breakdown products are capable of moving across the intestinal lining, into your bloodstream, and out to your body.

Your intestinal lining is designed for maximum absorption. It's not just a smooth layer of cells; it's densely folded into peaks and valleys, called villi (singular, *villus*) and looks a lot like a hair brush.

Weird Science

Each day, approximately 9 liters of fluid—all coming from your stomach, pancreas, and liver—empty into your small intestine. All of it, except 5 percent, is absorbed by your body across the intestinal lining.

Bio Buzz

Bile is a secretion of the liver. It consists of bile salts, bile pigments, cholesterol, and lecithin.

Weird Science

The colon, or large intestine, of a human adult is about 1.2 meters long when stretched out.

Through folding, the intestinal lining increases the surface area that's available for nutrient absorption. Each villus has a crown of microvilli. And each microvilli is a threadlike projection of the plasma membrane. Between the villi and the microvilli, your small intestine manages to get the maximum absorption from a minimum amount of space.

Each villus is a site for glucose and most amino acids to cross over the intestinal lining. Active transport moves nutrients across the plasma membranes of the villi. From there, they diffuse into extracellular fluid and enter the network of small blood vessels found inside each villus.

Once in the blood vessels, the nutrients enter your bloodstream and are carried out to other parts of your body. Mineral ions and water are also absorbed by your intestinal villi. By the time a meal has moved through the small intestine, everything that can be absorbed has been absorbed. Anything that's left moves into the large intestine, or colon.

The colon stores and concentrates feces, which are a mixture of water, bacteria, and undigested, or unabsorbed, material. Your colon starts off as a pouch, with your appendix a narrow projection from that pouch. From there, the colon winds its way through your abdominal cavity, snaking and turning upon itself, finally ending with a small tube called the rectum.

When waste material enters the rectum, the distension of the gut walls triggers a reflex to expel the feces from the body. Your nervous system controls this expulsion through a small sphincter muscle at the anus, or terminus of your digestive system. From mouth to anus, your food has been totally processed, all in one continuous gut tube.

Try It Yourself

Design a better diet for yourself and your family. The first step should be to increase your intake of fiber. The typical American diet doesn't include nearly enough bulk, or volume, of undigested fiber. Because there is insufficient volume in the colon, it takes feces longer to move through, which gives the bacteria (and sometimes carcinogens) a longer exposure to your gut. Scientists think this is one reason why Americans (and other "civilized" people) have a higher rate of colon cancer than population groups that eat a diet that's high in fiber.

You Are What You Eat: Nutrition

It appears that our early ancestors feasted on fruits, grains, and other fibrous plant materials, and ate meat only as an occasional treat. From these humble dietary beginnings, humans have moved into the consumption of saturated fats, sugars, salts, and very little fiber. We may be living longer than our ancestors did, but we are also suffering from diseases they never experienced: colon cancer, breast cancer, kidney stones, diabetes, and cardiovascular and circulatory diseases, most of which can be directly correlated to long-term changes in our diet.

Your body obtains all of its nutrients based on what you put in your mouth. It's no wonder that there's a direct link between your body's general health and *nutrition*.

Your body needs a steady supply of energy and materials from food in order to grow and maintain itself. And not just any old food will do. Your body needs specific foods, in specific amounts to obtain energy.

When it comes to food and your body, energy is measured in units called *calories*. When you read that a particular food—an apple, for example—has 100 calories, it sounds like there are 100 "things" in that apple, and you're going to consume them. Not true. A calorie isn't a "thing"; it's simply a unit of measurement. When you eat that apple, your body will receive the equivalent of 100 calories of energy.

To maintain a normal weight, your calorie intake must be balanced by energy output. In most adults, the energy input balances the energy output, and the body maintains a fairly steady weight over the years. If you take in more energy than you expend, your body will store that excess energy in the form of fat, or adipose tissue.

Bio Buzz

Nutrition is the process whereby an organism takes in and assimilates food or anything that nourishes.

Bio Buzz

A **calorie** is a unit of measurement for heat energy. One calorie is the amount of heat that's required to raise the temperature of 1 gram of water from 14.5° Celsius to 15.5° Celsius. Scientists use the term Calorie (with a capital "C") to express the amount of energy produced by food when it is oxidized in your body.

Your body needs several things to function properly. Complex carbohydrates (whole grains, beans, legumes, fruits, and vegetables) are your body's mainstay when it comes to energy. If you think back, it's carbohydrates that are most readily broken into glucose—and glucose is the primary source of energy for your brain, muscles, and all other body tissues.

Think About It!

There are some real benefits to maintaining a normal weight—and it's not just so you can look great at the beach. When you accumulate adipose tissue, your body has to produce way more capillaries than it ordinarily would to service these increased tissue masses. Consequently, blood volume drops, and your heart becomes stressed and must pump a lot harder to keep your blood circulating throughout your body—all of it.

Think About It!

Protein deficiencies are serious matters, especially when it comes to your children. A protein-deficient diet is most damaging in the young, when the brain and other tissues are developing rapidly. Unless a child gets enough protein just before and just after birth irreversible brain damage and mental retardation occur.

Sugar is a carbohydrate, but it's not a complex carbohydrate. It will add energy to your diet in the form of calories, but it won't provide any of the fiber and nutrients found in complex carbohydrates. Every year, the average American consumes as much as 128 pounds of sugar, which is more than 2 pounds a week. If that sounds ridiculous (oh you, with the healthy diet) take a close look at any label. Sucrose is frequently a common ingredient. Sucrose is just another name for sugar. You don't have to sprinkle it on your cereal to consume it.

Your body also needs protein. When proteins are digested, their amino acids become available for your body's use in its own protein construction. There are 20 common amino acids, and 8 of them are essential amino acids, meaning that your body cannot construct them; they have to be obtained directly from food.

Most animal proteins (meat, fish, dairy products, eggs, and so on) are "complete" in that they contain high amounts of essential amino acids. When you eat meat, you're getting all the amino acids your body needs. Plant proteins are "incomplete." They have all the essential amino acids required for body maintenance, but they don't have them in the right proportions. To make sure they get enough protein (the essential amino acids needed to maintain health), vegetarians need to eat certain combinations of different plants. In many world cultures, animal protein is a luxury used sparingly in the diet. These cultures have survived just fine by combining rice and beans, chili and cornbread, macaroni and cheese, tofu and rice, and lentils and wheat, to name a few.

Fats and lipids also play an important role in your body's health. Phospholipids and cholesterol are components of animal cell membranes. Fat deposits are used as energy reserves, but they also serve as cushions for many of your organs, such as your kidneys and eyes. You also need fats to help your body absorb fat-soluble vitamins.

The average American diet consists of about 40 percent fat, which most medical experts agree is too much. And there are fats, and there are fats. For health, you need to eat only 1 tablespoon of a polyunsaturated fat, such as olive oil or corn oil, a day. These types of oils contain linoleic acid, which is an essential fatty acid that your body cannot produce; you need to obtain it through your diet.

Saturated fats, such as butter and other animal fats, tend to raise the level of cholesterol in your body without providing any corresponding benefits. Your body needs cholesterol, but too much can have devastating consequences on your circulatory system.

Vitamins and minerals are the last vital components of a healthy diet. Normal metabolic activity relies on small amounts of organic substances called vitamins. Plant cells are able to synthesize vitamins, but animal cells have lost the ability to do so—or never had it. Either way, animals must obtain vitamins through their diet. Humans need at least 13 vitamins for optimum health.

Think About It!

Across the country, we are becoming a nation of eating disorders. For many people, nutrition is a subject fraught with emotional peril. Whatever the reason, find out more about the potentially fatal illnesses of anorexia (starving yourself) and bulimia (bingeing and purging) at http://eatingdisorders.miningco. com/health/eatingdisorders/ library/blanorexia.htm.

Your body also needs minerals, such as calcium, magnesium, and iron. About a dozen minerals are essential to good health and metabolic functioning.

Humans have one of the most intricate systems for obtaining and absorbing nutrients. From mouth to cellular activity, to elimination, your body is equipped to handle it all. Just send it down the hatch, and your body does the rest—for better or worse.

As you can see from what we've talked about in this chapter, ultimately, you really are what you eat. We've talked about how food enters your body and is digested and absorbed, and how good nutrition is the cornerstone of good health. In the next chapter, we'll be discussing your respiratory system—breathe deeply!

The Least You Need to Know

➤ The human digestive system is comprised of the mouth (oral cavity), pharynx, esophagus, stomach, small intestine, large intestine (colon), rectum, and anus. Salivary glands, as well as glands in the liver, pancreas, and gallbladder, aid in digestion.

➤ Your digestive system breaks down food molecules through mechanical, chemical, and hormone-assisted means.

➤ Starch digestion begins in the mouth, and protein digestion begins in the stomach. Digestion is completed in the small intestine, where nutrients are absorbed through villi in the lining and are delivered throughout your body by your circulatory and lymph systems.

➤ To maintain optimum health and a proper weight, your energy input must balance your energy output. Complex carbohydrates are the body's main source of energy, but protein, fats, vitamins, and minerals are also necessary for proper functioning of all your systems. Eight essential amino acids, a few fatty acids, and all vitamins and minerals can be obtained only through the diet.

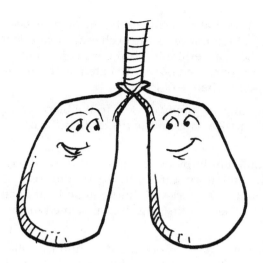

Chapter 14

The Respiratory System

In This Chapter

➤ In and out: the mechanism of breathing

➤ Gas exchange and transport

➤ Exercise and other oddities

Animals are the most active organisms on Earth. They're always on the go; even the lowly ameba is in constant motion. Where do animals get all this energy? A good diet surely helps, but that's not all it takes. Most of the energy that animals use for their functioning and daily activities comes from their aerobic metabolisms. Stated another way, the energy comes from respiration.

In this chapter, we're going to discuss how gases exchange within a body and fuel its activities. You don't have to be a land animal to respire; marine and aquatic animals also engage in respiration because gases can be found in water as well as air.

Always, Always, Always: Organization

All animals have some sort of respiratory system, and that's because they have an aerobic metabolism, which takes in oxygen, uses it, and then produces carbon dioxide waste. It doesn't matter where an animal lives—the top of Mt. Everest, the bottom of the Mississippi River, or the middle of the Pacific Ocean. It has an aerobic respiratory systems and uses oxygen to stay alive.

There are three types of respiration. External respiration is the exchange of gases between the atmosphere and the blood. Internal respiration is the exchange of gases between the blood and the cells of the body. Cellular respiration is the breakdown of glucose within the cell. In this chapter, we're concerned only with external and internal respiration because we're dealing with the exchange of gases.

The primary purpose of any respiratory system is to supply oxygen to the blood and to remove the resultant carbon dioxide that accumulates in the tissues that the oxygenated blood passes through.

The respiratory systems of animals are all different in many respects, but they all rely upon a thin, moist layer of epithelium (the respiratory surface) that gases diffuse across. The surface has to remain moist at all times; gases will diffuse in and out of an animal only if they are first dissolved in some type of fluid.

Every respiratory system is organized around this diffusion of gases. Oxygen, carbon dioxide, and other gases diffuse pressure gradients. That is, the more concentrated the gas molecules are outside your body, the greater force is available to drive those same individual molecules inside your body. Likewise, the system works in reverse: A lower concentration of molecules outside your body will pull gas molecules out of your body.

Oxygen and carbon dioxide aren't the only gases in the atmosphere that your body uses. At sea level, the air we breathe is about 78 percent nitrogen, 21 percent oxygen, 0.04 percent carbon dioxide, and 0.96 percent other gases. Each gas exerts a partial influence on the pressure that operates your respiratory system.

Most gases, particularly oxygen and carbon dioxide, tend to diffuse from areas of high pressure to areas of low pressure. All respiratory systems take advantage of this high-to-low tendency of gases by working with the partial pressure gradients that exist between the internal environment (the inside of an animal's body) and the external environment.

How does an animal's body determine just how much oxygen (or any other gas) is diffusing across the moist respiratory surface at any given time? Do these gases just flood in? According to *Fick's Law*, the amount of gas diffusing across the membrane depends on two things: the surface area of the membrane, and the difference in the partial pressure on either side of it.

Try It Yourself

The next time you take a breath, think about this: It's the tendency of gases to move from high-pressure to low-pressure areas that enables you to breathe in and out. The air in your lungs is going from high to low areas.

Bio Buzz

Fick's Law states that at any given time, the diffusion of gases across the respiratory membrane depends upon the amount of surface area of the membrane and the partial pressure across it.

To explain this rather obscure law, we need to talk about both these things, membrane surface and partial pressure. Remember that gas can diffuse only across a moist respiratory, or membrane, surface. It follows, then, that the greater the surface area of the membrane, the more gas will diffuse across it at any given time.

As for partial pressure, not every gas exerts the same pressure. If you fill a flat tire at a sea-level beach—say, San Diego—you're filling that tire with about 78 percent nitrogen, 21 percent oxygen, 0.04 percent carbon dioxide, and about 0.96 percent other gases. These ratios are true for any dry, sea-level air. Out of the total pressure being exerted against that filled tire, oxygen is exerting only part of it. Partial pressure refers to the contribution a particular gas is making to the total exerted pressure. When it comes to respiration (and Fick's Law, in particular), the greater the partial pressure gradient of a particular gas, the faster it will diffuse across the respiratory membrane.

If a larger membrane surface promotes greater gas diffusion (and all animals need those gases to function), then why don't all animals have huge respiratory membrane surfaces, respiratory surfaces that cover their whole body? After all, isn't bigger better? This pattern doesn't work in large animals because if the membrane surface just keeps expanding (like a balloon), in no time at all, the animal's volume (in this case, the air inside the balloon) is much greater than the animal's surface area (the balloon). There's too much volume to service; the diffused gas can't reach the center. Without the adaptation of a true respiratory system, large, complex animals couldn't possibly service their internal functions.

Animals without true respiratory organs (such as flatworms and roundworms) have flattened, tube-like bodies because their internal cells have to be close to the respiratory surface. Because they don't have actual respiratory organs, they need this type of body design in order to survive. But this type of system works only in small animals without a lot of volume. Otherwise, diffusion over the surface area by itself doesn't get an animal very far.

Besides body shape, animals have acquired other adaptations to overcome constraints imposed by their particular respiratory systems. When a fish fans its gills, it's actively ventilating its body surface by stirring up the water so that it has plenty of the necessary gases (oxygen) in the immediate

Weird Science

What about marine mammals? How do they stay underwater so long? Seals, for example, have greater blood volume and more hemoglobin than humans do. When a seal starts to dive, the air in its lungs converts into their bloodstream and binds to the hemoglobin, giving it a greater supply than a human merely holding his breath. Courtesy of gas transport, when marine animals dive, their blood stream acts almost like our lungs do: It keeps delivering plenty of oxygen to all parts of their bodies. Seals also have more myoglobin in their muscles, which, if you recall, also binds with oxygen. So, although seals aren't technically "breathing," they're still receiving plenty of oxygen while under water.

area. Likewise, the microvilli on the collar cells of sponges do more than just sweep in food. They also ventilate the sponge's surface and help improve diffusion rates.

Gas Diffusion in Aquatic Environments

Water contains oxygen; it's not called H_2O for nothing. That "O" stands for oxygen. If we humans tried to obtain our oxygen from water, we'd die, but other animals are able to live by getting their entire oxygen supply from water. However, those that do have a harder time of it than terrestrial animals. That's because, liter for liter, water contains less oxygen than air. A liter of water that is completely saturated with dissolved oxygen still contains only about 5 percent as much oxygen as a liter of air does.

There are also differences between aquatic environments. Freshwater holds more oxygen than saltwater, and the saltier the water, the less oxygen. The Great Salt Lake and the Dead Sea, for instance, have far less oxygen in their waters than the Atlantic Ocean does. In addition, the oxygen in still or stagnant water, such as you might find in a small pond, becomes depleted much faster than the oxygen in larger bodies of moving water. Without circulation, the oxygen in the water gets used up faster than it gets replenished. The absence of sunshine also influences the production of oxygen. The less sunlight, the fewer photosynthetic organisms in the water, and the less available oxygen.

Weird Science

For terrestrial animals, there's such a thing as too much when it comes to moistness. Earthworms obtain their oxygen from the soil, and when rain floods the spaces between the particles of soil, oxygen levels drop and carbon dioxide levels rise. That's why you often see earthworms wiggling around on the grass after a thunderstorm: They're trying to get oxygen.

Gas Diffusion in Terrestrial Environments

Land animals have a much easier time of it when it comes to getting oxygen. They're able to get more oxygen into their systems by using less energy. But there are some pitfalls. For instance, although air has more oxygen than water, it also poses a greater threat to respiratory systems. If the respiratory surface becomes dry, it won't work. The membranes dry out, stick together, and cannot allow the exchange of gases.

Bio Buzz

In an **integumentary exchange** respiratory system, oxygen and carbon dioxide simply diffuse across the thin moist layer of surface skin (epidermal) cells.

Organization and Diversity

Several types of respiratory systems are found in animals. The main ones are integumentary, tracheas, gills, and lungs. *Integumentary exchange* systems, which involve the simple diffusion of oxygen and carbon dioxide across the skin, work only in relatively small, relatively sedentary animals. Most annelids, some arthropods, and nudibranchs have integumentary exchange respiratory systems, as do most frogs and amphibians.

For water-dwelling animals, the exterior environment keeps the surface epidermal cells moist, allowing the exchange of gases. Terrestrial (or semiterrestrial) animals, such as amphibians, have to rely on mucus secretion to keep the respiratory surface (their "skin") moist.

Insects and spiders have respiratory systems consisting of chitin-reinforced tracheas, which are tubes that connect the body to the source of oxygen. (Chitin is the same tough, horny material that covers the outside of an insect's body, basically its exoskeleton material.) In insects, the tracheas branch throughout the body, providing a self-contained system of gas conduction and exchange. No assistance from the circulatory system is required; frequently, a small "lid" covers the surface opening of the trachea and prevents its interior from drying out. Tracheas branch again and again, finally ending in dead-end tips filled with fluid, which provides the necessary moisture for gas exchange. Tracheas frequently dead-end against cells that have high oxygen requirements, such as muscle cells. Humans also have tracheas, and we'll discuss them a little later.

Fish and the larval forms of some amphibians respire through *gills*. The most commonly known gills are those of adult fish. Their internal gills are composed of separated layers of thin tissue (epidermis) that extend from the back of the fish's mouth out to the surface, just in back of the fish's head. Water enters through the mouth, travels down the pharynx, flows out across the gills, and ends up back in the external environment. Oxygenated water moves over the gills, and blood circulates through them. The blood picks up the oxygen and delivers it to other parts of the body. Fish obtain about 80 to 90 percent of the oxygen they use in this manner.

The respiratory system of reptiles, birds, and mammals involve paired *lungs*. Lungs evolved more than 450 million years ago and first showed up, believe it or not, in fish. As near as scientists can tell, the development of lungs in fish allowed them to

Weird Science

Ever noticed how a bee that's buzzing around and gathering pollen will frequently stop and pump the segments of its abdomen back and forth? Flying is energy-intensive work. The bee is forcing air in and out of its trachea system, delivering fresh oxygen to the muscle cells in its wings.

Bio Buzz

A **gill** is a respiratory organ comprised of a thin, moist layer of epidermis. The larval forms of some fish and some amphibians have external gills. Adult fish have internal gills.

Bio Buzz

A **lung** is an internal respiratory surface, shaped like a sac or a cavity.

Weird Science

Your nervous system is what ultimately controls your breathing. Neurons sense oxygen and carbon dioxide levels throughout your body and respond by adjusting the contractions of your diaphragm and the muscles in your rib cage, thereby controlling your inspiration and expiration.

Bio Buzz

A **bronchus** is one of two branches of the lower part of the human trachea. Each bronchus leads to a lung.

obtain more oxygen from an environment that was oxygen-deficient. In some fish lineages, the lung still exists in the form of a swim bladder. Even today, African lungfish have both gills *and* lungs; if something prohibits their ability to gulp air at the surface, they drown. Although amphibians rely upon integumentary exchange for most of their respiration, most supplement oxygen intake with a pair of small lungs. Lungs are most important in vertebrate respiration.

In all lungs—mammal, bird, or reptile—airway tubes carry gases to and from one side of the respiratory surface. The circulatory system, or blood vessels, carry gases to and from the other side of the respiratory membrane. No matter what type of respiratory system an animal has—integumentary exchange, trachea, gill, or lung—all are highly organized and serve the same purpose: to diffuse gases and to help an oxygen-dependent metabolism survive.

In and Out: The Mechanism of Breathing

Having discussed different animal respiratory systems, it's time to get back to looking at the human respiratory system.

Breathing is the process of inspiring and expiring air. Hopefully, for your own health and survival, the inspired air is oxygen-rich, and the expired air is rich in carbon dioxide. But as far as the mechanics of breathing go, inspiration is inspiration, and expiration is expiration. If things other than oxygen are in the mix, it's your body's job to deal with it. Your body will breathe, come what may.

You breathe both voluntarily and involuntarily. For instance, you can consciously control the amount of air you inspire by deliberately holding your breath or gulping in extra air, at least for a little while. But you don't *have* to be consciously aware of every breath you take. You can become involved in a fascinating conversation or movie and never give your breathing a backward glance. You can even go to sleep and be fairly confident that your lungs will keep doing their job.

The mechanics of breathing involve your diaphragm, abdomen, and ribs. Each phase of breathing—inspiration and expiration—requires the coordinated interaction of these three areas. When you take a deep breath (the process of inspiring), you can feel your chest expand. As your muscles contract, your diaphragm flattens and pushes down on your abdomen. When your diaphragm flattens and your rib cage pushes out, the size of your thoracic cavity increases.

This increase in size causes the air pressure in your thoracic cavity to drop; it becomes lower than the air pressure outside your body. As a consequence, air from the atmosphere moves into your lungs and equalizes the pressure. Expiration is the same process, only in reverse. The air pressure inside the thoracic cavity becomes greater than the air pressure outside your body, so your lungs expel the air to equalize the pressure.

Inspiration starts at your mouth and nose. When you breathe in, air filters through a network of small hairs in your nose (we're going to assume that you don't have a cold and that your nose is working just fine) and passes into your nasal cavity, which is located just above the roof of your mouth. The nasal cavity has mucous membranes that moisten and warm the air, which helps prevent damage to delicate lung tissues. The moistened, filtered air then moves into the pharynx and into your trachea.

Your trachea is a cartilaginous windpipe. In structure and function, it's not much different from the tracheas found in insects. It's tough and flexible, and it transports air. Your trachea is about 12 centimeters long and is lined with ciliated cells. At the upper end of your trachea is your larynx, or voice box. Sounds are produced (singing, talking, incoherent grunting) when air is forced past two ligaments, your vocal chords, that stretch across the interior of the larynx. The lower end of your trachea branches into two bronchi (singular, *bronchus*), each of which leads to a lung.

Once in the lungs, the bronchi start to resemble the trachea of insects: They branch and branch and branch some more, creating a weblike network throughout each lung. Starting with a main tube coming off the trachea, each bronchus repeatedly branches into smaller and smaller tubes called bronchioles, each of which ends in a cluster of tiny air sacs called alveoli. Bronchi walls consist of smooth muscle, cartilage, ciliated cells, and mucus. The very smallest tubes at the end are lined with cilia and mucus.

Think About It!

How much do you breathe? One respiratory cycle equals one completed inspiration plus one completed expiration. At rest, most humans breathe at a rate of 16 to 24 cycles per minute.

Think About It!

Ever swallow the wrong way? If you recall, your pharynx is also part of your digestive system. When you're not eating and swallowing, your epiglottis remains open so that air can pass through the pharynx and into your trachea.

183

A network of tiny blood vessels surrounds each alveoli, and this is where gas exchange takes place in your body, between the alveoli and blood vessels. Each human lung contains more than 300 million alveoli, each one measuring only 0.1 millimeter to 0.2 millimeter in diameter. Taken together, your alveoli provide a total surface area of nearly 70 square meters for gas exchange in both lungs.

Gas Exchange and Transport

The actual mechanics of respiration gets the air into your body, but without gas exchange, it's no different than repeatedly blowing up a balloon and deflating it. During gas exchange, vital, life-giving oxygen dissolves across a fluid surface and is transported out to waiting cells, which then use it to fuel cellular activity. Each alveoli (all 600 million of them) is only a single layer of epithelial cells, surrounded by a thin membrane. Each alveoli is a lot like a teeny balloon, only a single layer of cells thick. The only thing separating the alveoli from the blood vessels in your lungs is a thin film of *interstitial fluid*. Interstitial fluid is fluid that fills in the spaces between cells and tissues in complex animals. The interstitial fluid provides the moisture needed to dissolve gases, and it's only a thin film, which makes it easy for gases to rapidly dissolve across it and into the blood vessels.

Bio Buzz

Interstitus means "to stand in the middle of something." **Interstitial fluid** is the extra-cellular fluid that occupies the spaces between cells and tissues.

Your body doesn't need to expend any energy for this diffusion to take place. By itself, passive diffusion moves oxygen across the boundary and into your circulatory system, and moves carbon dioxide in the reverse direction. The partial pressure gradients of oxygen and carbon dioxide allow this to happen.

Although diffusion happens effortlessly, your blood can transport only so much oxygen and carbon dioxide in dissolved form. Humans and other large-bodied animals need a much better, more efficient transport system in order to meet their needs. For us, it's hemoglobin, an iron-containing protein found in blood, that makes the difference by providing us with a transport system for both oxygen and carbon dioxide. Hemoglobin transports approximately 97 percent of the oxygen your tissues use.

Each hemoglobin molecule in your bloodstream can carry up to four molecules of oxygen. The oxygen molecules bind with the central iron molecule found in hemoglobin. When the oxygenated blood reaches your tissues and cells, the hemoglobin releases the oxygen, and it diffuses out through capillaries into the surrounding cells.

Hemoglobin transports in both directions. After cells use the oxygen, they excrete the waste products of carbon dioxide into your bloodstream. About 8 percent of the carbon dioxide dissolves into your blood, and 25 percent binds onto hemoglobin molecules and make the return trip back to your alveoli, where they are expired.

What about the other 67 percent? Thanks to the innate genius of your body's design, your body converts the remaining carbon dioxide molecules into bicarbonate ions. Bicarbonate ions are ions with an electrical charge, carrying a more transportable form of carbon dioxide. And guess what? Bicarbonate ions dissolve beautifully in your bloodstream. When your bloodstream naturally cycles back to your lungs, a reverse reaction takes place. The carbon dioxide diffuses through your alveoli capillaries and is exhaled into the atmosphere.

Weird Science

One of the perils of cigarette smoking is that the smoke itself contains a high concentration of carbon monoxide. Unfortunately, hemoglobin can't differentiate between the carbon monoxide that is cellular waste and the carbon monoxide that comes from cigarette smoke. The carbon monoxide of cigarette smoke binds more easily to hemoglobin than your own respiratory carbon monoxide does. The carbon monoxide from cigarette smoke will hog most of the binding sites. Therefore, when you smoke, your hemoglobin does double duty (it's scurrying to get rid of both types of carbon monoxide), and, consequently, your heart must also work harder to pump oxygen-rich blood to your cells. This is why smoking affects your heart as well as your lungs.

Exercise and Other Oddities

Your respiratory system is not static. At any given moment, external and internal forces can cause it to make adjustments. Exercise and increased or decreased air pressure are just a few of the things that cause your body's respiratory system to alter.

Exercise: Go, Go, Go!

When you exercise, everything springs into action: nervous system, muscles, heart and blood vessels, and yes, your respiratory system. Although your nervous system controls the rate of your breathing, local controls also come into play. When your heart is pounding and you're not breathing deeply enough, carbon dioxide starts to build up in the alveoli. The pressure increase pushes against the smooth muscle walls of the bronchioles, causing them to dilate, which matches up the rates of air and blood flow.

Because exercise increases the rate of air and blood flow, it's also a good way to increase your total lung capacity. Taken together, the maximum amount of air both your lungs can hold is somewhere between 5 and 6 liters of air. But humans never inhale or exhale this amount. Breathing normally, you inhale only about 0.5 liters of air. Even during deep breathing, you never entirely fill or empty your lungs.

But, as you can see, when you exercise strenuously, you're exercising your lungs as well as your muscles. Through this type of conditioning, your lungs get used to taking in more air (and exhaling it), and you can get to the point where you take in 4.5 liters of air in a single breath.

Try It Yourself

Over the centuries, science has advanced because people thought and pondered things of interest. Many questions are not immediately answerable, and some may never be. Put on your thinking cap and see if you can advance the scientific cause.

Before a polio vaccine was invented, this virus affected countless people, paralyzing their muscles and the nerves that controlled them. Many people who contracted polio died because they couldn't breathe. Based on what you've learned about the respiratory system, what are some ways the polio virus might affect this system?

As a possible consequence of aggressive lawsuits against the tobacco industry, many cigarette manufacturers are now urging their patrons to "smoke responsibly." Based on what you know of the human respiratory system, is there such a thing, biologically and healthfully speaking? Is there such a thing as "responsible smoking" when it comes to a child, a nonsmoking spouse, and patrons of public places? What is "responsible"?

Respiratory systems can efficiently handle most of the problems thrown their way, as long as that one key element is present: oxygen. It's the key to life.

We've discussed some complicated subjects in this chapter, namely how gases manage to move across and through fluid, fueling bodily functions in the process. In the next chapter, we'll be discussing the circulatory system. Because hemoglobin and your bloodstream are such an integral part of respiration and gas transport, these two systems really go hand in hand, as you'll see.

Think About It!

If you got "the bends" in the ol' days, chances are that you would die. But nowadays if you find yourself in that situation, you'll be put into a decompression chamber. Some diving boats keep them on board, and many hospitals have them. A decompression chamber applies pressure (again) to your body, compressing the nitrogen back into your bloodstream. Then the pressure is lessened very gradually. It will take several hours, but eventually you'll be as good as new.

The Least You Need to Know

➤ Animals are the most active of all organisms. Most of their energy comes from an aerobic metabolism that uses oxygen and produces carbon dioxide wastes.

➤ Respiration is the process whereby animals move oxygen from the external environment into their internal environment and give up carbon dioxide in return.

➤ All respiratory systems use the tendency of gases to diffuse to a lower pressure gradient. Diffusion is passive and does not require cellular energy.

➤ Gases can diffuse only in fluid. All respiratory systems rely upon a respiratory surface, a thin, moist, membranous surface that gases can readily diffuse across.

➤ In humans, oxygen molecules bind to hemoglobin in the bloodstream and are then transported throughout the body as fuel for cellular activity. Carbon dioxide waste is transported out through the same method, in reverse.

The Circulatory System

Although your nervous system is the command center for your body, your circulatory system delivers the goods. Your circulatory system delivers everything from nutrients for cellular energy to warrior white cells prepared to do battle with hostile invaders such as viruses and bacteria. If you get a cut or a scrape, it's your circulatory system that comes to the rescue, delivering substances that close the gap. Your circulatory system even helps control your body's temperature.

All animals have some sort of circulatory system. Circulatory systems are the major transport highways in any organism. In this chapter, we'll explore the components of this complex, efficient system in humans and other mammals.

Always, Always, Always: Organization

What if an earthquake occurred in your area, closing down all highways? Grocery trucks couldn't get into the area, and supplies were running low. Refuse started piling up because garbage trucks couldn't get out of the area.

Your circulatory system is just like a highway. Nearly everything in your body uses this highway to get around, including: nutrients, waste products, and cells that fight off disease. If your circulatory system were disrupted, every cell in your body would face a similar predicament: No nutrients could get in, and no garbage could get out. Let's take a look.

The Basics

Most invertebrates and all vertebrates have circulatory systems composed of three basic components:

➤ **Blood**—A liquid connective tissue composed of water, solutes, and other formed items such as platelets and blood cells

➤ **Heart**—A muscular pumping organ that keeps blood moving throughout the body

➤ **Blood vessels**—Different-sized tubes through which blood is transported to all parts of the animal's body

As with everything else, there's a lot of variety among animal circulatory systems. Most of these systems don't appear to be even remotely related to each other. For instance, the blood of a grasshopper is markedly different than your own blood. Yet you and the grasshopper each have a circulatory system that shares the same three basic traits.

Try It Yourself

Believe it or not, you can tell how hot or cold it is outside by listening to crickets and grasshoppers. Relatively speaking, insects have small bodies, and most of their blood is close to the surface. Consequently, their circulatory systems are easily affected by external temperatures, so much so that crickets and grasshoppers will even tell you the outside temperature. Count the chirps per minute, subtract 19, divide by 2, and then add 60. That's it—probably the temperature in Fahrenheit of a balmy summer night. (How this whole thing works is a mystery, but what's more mysterious is who thought about it or came up with the initial equation!)

You each have liquid connective tissue (blood), a pumping mechanism (a heart) that keeps the liquid connective tissue circulating, and a network (blood vessels) that delivers the liquid connective tissue to every cell in the body.

Animal circulatory systems are either open or closed. Arthropods and most mollusks have open circulatory systems. Most other animals have closed circulatory systems.

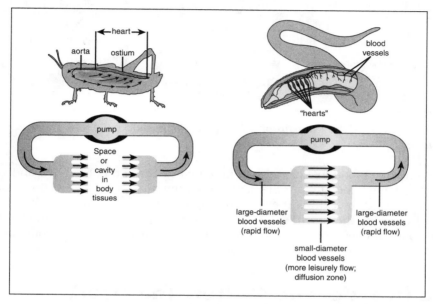

All arthropods and most mollusks have open circulatory systems. Other animals have closed circulatory systems.

In an open circulatory system, the heart pumps blood into an open space, or body cavity, where the blood mingles with the body tissues. Open-ended tubes, which lead back to the heart, line the body cavity. Because the blood has nowhere else to go, it eventually ends up in these tubes and then back at the heart. The heart pumps it out again, and the process starts all over.

Open circulatory systems are effective for small bodies, such as those of insects, but they don't work in larger animals. Most animals, from earthworms to humans, have closed circulatory systems. In a closed circulatory system, the heart and the blood vessels are continuously connected, and the blood is contained within this closed loop. The volume of blood that flows through the system is always equal to the volume of blood being returned to the heart. The heart constantly pumps whatever blood it receives. Even among such diverse groups as fishes, amphibians, and mammals, closed circulatory systems are remarkably alike.

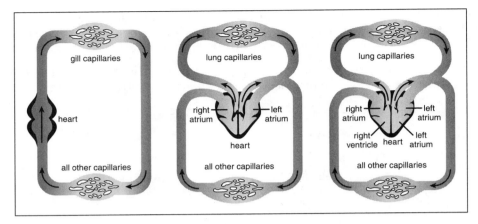

The closed circulatory systems of fishes, amphibians, and mammals are very similar.

The volume of blood in a closed system remains constant, but the rate and volume of the flow changes along the route depending primarily on the size of the vessel. Blood starts its journey by exiting the heart through relatively wide tubes, referred to as arteries and veins. By the time it reaches the outermost parts of the system, blood is traveling through microscopic tubes, or capillaries. In humans, the final capillaries blood travels through are so small that red blood cells can fit through only one at a time.

At the Heart of the Matter: The Heart

At the center of any circulatory system is the heart, a muscular mechanism that pumps blood throughout the body. A human heart like yours is a durable pump, consisting of cardiac muscle, nervous tissue, and connective tissue. Your heart is located in your thoracic cavity, just behind your sternum and between your lungs. Most of your heart is composed of cardiac muscle tissue that is surrounded by a tough protective sac, called the pericardium. The pericardium secretes a fluid that reduces friction when the heart beats.

The septum divides your heart lengthwise into two halves. Each half of your heart serves a different function. The right side pumps blood into your lungs, and the left side pumps blood into other parts of the body. Each half of your heart is further divided into two chambers. The atrium is the upper chamber, and the ventricle is the lower chamber. Flaps of membrane, your AV valves, (short for "atrium and ventricle") separate the atrium and ventricle, acting like a one-way valve.

Weird Science

During the average 75–year life span of a human, the human heart will beat approximately 2.5 billion times, resting only a brief moment between beats. The adult human heart is about the size of a fist.

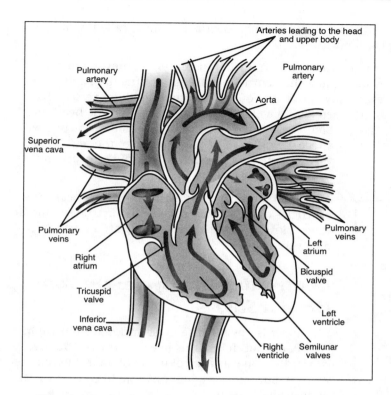

Arteries leading to the head and upper body

Pulmonary artery

Pulmonary artery

Aorta

Superior vena cava

Pulmonary veins

Right atrium

Tricuspid valve

Inferior vena cava

Left atrium

Bicuspid valve

Pulmonary veins

Left ventricle

Right ventricle

Semilunar valves

The human heart is divided lengthwise into two halves. Each half is composed of an upper chamber (the atrium) and a lower chamber (the ventricle). Arteries and veins deliver blood to and from the heart.

Your heart also has other valves, the semilunar valves, located between the ventricles and the arteries that lead away from them. Every time your heart beats, its valves open and close in a synchronized movement that prevents back flow and keeps the blood moving in one direction.

The inner chambers of your heart have a smooth lining, called the endocardium. The endocardium is comprised of connective tissue and a single layer of epithelial cells (cells that comprise the surface layer). Besides lining your heart, this epithelial layer (called the endothelium, when located in your heart) also lines the inside of blood vessels.

The Cardiac Cycle

Each time your heart beats, the four chambers go through a sequence of muscle contraction and relaxation called the *cardiac cycle*. As your atria relax, they fill with blood. As they fill, the pressure builds, which forces the AV valves to open and

Bio Buzz

The **cardiac cycle** is a sequence of contraction and relaxation of the heart muscles. The contraction phase is referred to as **systole,** and the relaxation phase is called **diastole.**

blood to flow into the ventricles. The atria then contract, a phase called *systole*, which completely fills the ventricles.

The filled ventricles start to contract, which further increases the pressure inside them, and the AV valves snap shut. The ventricles continue contracting, and pressure inside the ventricles rises sharply above that of the surrounding blood vessels that lead away from the heart. (Remember, in a closed circulatory system, the heart and blood vessels are all connected; what affects one will affect the other.) The pressure forces the semilunar valves open, and blood flows out of the heart and into your blood vessels.

After the blood has been ejected from the heart, the ventricles relax, a phase called *diastole*. The semilunar valves close, and the already-filling atria start the entire cycle again.

Controlling the Cycle: Do You Have a Say?

Your heart is possibly the most important muscle in your body, but did nature give you any control over it? Hardly. You can control many of your muscular responses, but rarely are humans able to control the actions of their hearts.

As a muscle, your heart has little in common with your skeletal muscles. Rather than individual cells that attach to your bones, the cells of your cardiac muscle tissue branch out and connect with each other at their endings. All the cells communicate through the plasma membrane that abuts the joined cells. Each time your heart beats, a signal calling for "contraction" spreads out so rapidly through the cellular network that all the cardiac muscle cells respond as if they were one unit.

In the case of your skeletal muscles, your nervous system delivers the signals that make them contract. But when it comes to your cardiac muscle, your nervous system can only *adjust* the strength and the rate of the contractions, it can't control the actual contractions. Your cardiac conduction system does that.

Conducting cells in the SA node (which is short for *sinoatrial*), located where the major veins enter the right atrium, triggers the contraction. The SA node

Think About It!

If you ever feel unconnected to your internal functions, take a minute and listen to your heart. The familiar "lub–dup" sound of your heartbeat is actually your valves opening and closing. The "lub" sound is the AV valve closing and ventricles contracting. The "dup" sound is the semilunar valves closing and the ventricles relaxing.

Think About It!

For a fun sci-fi journey through your body, rent a copy of the video *Fantastic Voyage*, an old movie about a group of scientists and doctors that shrink themselves (and their submarine–like vessel) and journey through the human body.

generates continuous waves of excitement, usually 70 to 80 every minute. Each wave spreads out over the atria and causes them to contract. Within a tenth of a second, each wave travels over the atria and reaches the AV node (short for *atrioventricular*). The AV node acts somewhat like a transformer station: The signal slows down a bit when it reaches the AV node. This ever-so-slight delay gives the atria time to finish contracting before the next wave moves on to the ventricles and causes them to contract.

Although all the cells in this system are self-excitatory, the SA node starts the firing of the nerve impulses at the highest frequency. The SA node is also the first place to reach threshold in the cycle and start the contraction. Consequently, the SA node is referred to as the cardiac pacemaker. Its rhythmic firing of nerve impulses creates the normal human heartbeat.

The Delivery System: Blood Vessels

In the cardiovascular system of all vertebrates, the heart pumps blood into large blood vessels called *arteries*. From there, the blood flows into smaller, muscular arterioles, which in turn branch into a network of tiny capillaries. From the capillaries, the blood flows into tiny venules and then into large *veins* that return the blood to the heart.

Scientists divide your cardiovascular system into two different circuits: the pulmonary circuit and the systemic circuit. Pulmonary circulation starts when blood flows from the right ventricle, through the semilunar valve, and into the pulmonary artery. At that point, the blood is on its way to your lungs to pick up oxygen. The pulmonary artery is the only artery in your body that carries deoxygenated blood; all other arteries are carrying oxygenated blood.

At the lungs, the pulmonary artery divides into two smaller arteries, each one leading to a lung. Once in the lung, each artery branches into several smaller arterioles and continues branching until it has created a network of capillaries. The blood in these capillaries both releases the waste carbon dioxide it has picked up from cellular activities and absorbs oxygen (through gas diffusion when it

Bio Buzz

Arteries carry oxygenated blood. **Veins** carry deoxygenated blood.

Weird Science

The word *cardiovascular* comes from the Greek word *card*, which means "heart," and the Latin word *vasculum*, which means "vessel."

comes in contact with the alveoli, the smallest air sacs in your lungs). The oxygenated blood flows into venules, which in turn merge into larger pulmonary veins and then into the left atrium of the heart. The circuit the blood travels from the heart to the rest of the body (apart from the lungs) is called the systemic circuit.

The systemic circuit starts when blood from the left ventricle flows through the semilunar valve and into the aorta, which is your body's largest artery. Once in the aorta, the blood is on its way to all parts of your body (except your lungs), thanks to a branching network of arteries, arterioles, and capillaries. The systemic circuit works just like the pulmonary circuit, except that the blood leaving the heart is oxygenated.

By the time the blood reaches the last capillary network, it will have left all its oxygen behind for the cells and picked up carbon dioxide in return. Then, moving through a series of venules and veins, it will wend its way back to the heart and head out to your lungs, to pick up oxygen, return to your heart, and start out all over again.

That's the simple version. Actually, your systemic circuit is a little more complex than that because your body is complex. An earthworm, you are not. Scientists divide your systemic circuit into three other subsystems: coronary circulation, renal circulation, and hepatic portal circulation.

Think About It!

Do your heart some good by watching your diet. Atherosclerosis is a disease that's characterized by the build-up of fatty substances on the walls of the coronary arteries. If the blockage shuts off the flow of blood to the heart, a heart attack will result. The more fatty foods you eat, the more likely it is that you'll develop atherosclerosis. Your family history of the disease and the amount of exercise you get are two other factors that influence its development.

Coronary Circulation

Like any other muscle in your body, your heart needs a steady supply of blood to keep functioning. But most of the muscles in your heart are not serviced by the blood that's flowing through its chambers. Your heart has its own circulatory network, with two main arteries leading into an extensive capillary system within your heart. Coronary circulation is one of the subsystems of your systemic circulation.

Two large coronary arteries branch off from your aorta and transport blood into arterioles that branch off and penetrate the tissue of your heart. The blood then returns into the right atrium through a large vein, referred to as the coronary sinus.

With all the blood traveling in and out of your heart, you might think that it wouldn't need its own arteries. But every cell needs oxygen (delivered by your bloodstream), and the muscle of your heart is very thick. Your heart cells need oxygen so much that if a coronary artery becomes clogged (shutting down the supply of oxygen to your heart cells) those muscle cells will die.

Renal Circulation

Renal circulation is the systemic system that moves blood into your kidneys, where waste products are excreted, and back again to your heart. Two renal arteries branch off the aorta, each going to a kidney. The nitrogenous wastes (along with carbon dioxide, these are waste products of your cellular functions—these wastes diffuse into your blood stream the same way carbon dioxide or oxygen does) are filtered out by the kidneys, and the blood is then returned to the heart by renal veins.

Arteries and Veins: The Main Highways

Your blood vessels and heart are connected, so it's not surprising that both share some common structural traits. Like your heart, arteries and veins are constructed in layers: endothelium (the inner layer), smooth muscle (the middle layer), and connective tissue (the outermost layer). This layered structure gives your blood vessels the strength and flexibility they need to stretch when blood leaves the heart and enters the system under pressure.

The layer of smooth muscle is thicker and stronger in your arteries than in your veins. Any ideas why? When oxygenated blood leaves your heart to make its way to the outposts of your body, it starts out as a large flow under pressure (thanks to your heart, that marvelous pump). As the arteries turn into arterioles, the pressure decreases because the blood volume is no longer concentrated in one place; it's working its way from one large tube (the artery) and is spreading out through smaller tubes. The smooth muscle in your arteries is thicker and stronger than the tissue in veins because your arteries deliver the oxygen, and your cells want that oxygen right now! As the blood gets farther from the heart and the pressure decreases, the muscles in your arteries help keep it moving along. By the time it reaches the capillaries, the pressure is almost nonexistent, but your artery muscles are still working. Red blood cells move in single file

Bio Buzz

Human **blood** is a liquid connective tissue comprised of water, solutes, and uniquely formed elements, such as platelets and blood cells.

Think About It!

When you're feeling a bit woozy or shaky, there might not be enough nutrients in your bloodstream. To see how fast your bloodstream works, eat a banana. Your digestive system will go to work almost immediately, and your bloodstream will start delivering equally fast; you should be feeling better in less than five minutes. (You want to eat a banana rather than, say, steak, because it's pure carbohydrate and goes out to your cells almost immediately; steak has to digested before it can be delivered.)

through the capillaries, allowing plenty of time for diffusion to take place. Your capillaries are where body cells absorb oxygen and nutrients from the blood in exchange for carbon dioxide.

When the capillaries merge into venules (the smallest kind of veins and then into the larger veins, the blood starts the return trip to the heart. The smooth muscle in your veins is thinner than that in your arteries because there's no real rush; the veins don't deliver vital nutrients and oxygen. Veins move the depleted blood along back to the heart, and then the aorta pumps it into the lungs to pick up more oxygen.

The Blood of Life: Blood

Your heart pumps it and your blood vessels move it, but what exactly is it that this system moves? Blood is such a common substance (after all, all animals have it) that few people stop to ponder what it really is. Blood is blood, right? Yes and no. Blood certainly stands by itself as a unique substance. But *blood* is actually a tissue; a liquid connective tissue.

Blood serves several functions. It carries nutrients and oxygen to your cells, and it carries wastes away from them. It distributes hormones throughout your body. It also serves as a highway for your body's police force, the fighting white cells of your immune system. In mammals and birds, blood helps maintain a constant body temperature by carrying heat away from areas of high metabolic activity (such as your skeletal muscles during exercise) to the surface areas of the skin, where the heat is dispersed.

Think About It!

When you give blood, your plasma is another gift of life that you offer. By itself, without any blood solids, plasma contains so many essential items that giving someone just plasma can save that person's life in a trauma situation. Unlike other blood components, plasma is not typed, so everyone can use it, and it's also easier to store.

Your blood can be divided into two main components: plasma and blood solids. Plasma makes up about 55 percent of your total volume, and blood solids make up the remaining 45 percent.

Plasma

Plasma is a sticky, pale-yellow fluid that's made up of roughly 90 percent water. Plasma is the transport carrier for the bulk of the nourishment that goes out to cells. Dissolved substances, such as vitamins, minerals, amino acids, glucose, and other things, are absorbed through the walls of the digestive system and are transported by the plasma out to cells, where they're absorbed and used to fuel cellular functions. Plasma also transports most wastes back to the kidneys and lungs so that they can be removed from the body.

Various proteins can also be found in plasma. Some, such as albumin, influence the distribution of water between the blood and the interstitial fluid in your body, so plasma actually influences your blood's fluid volume. Albumin represents about 60 percent of your body's plasma proteins. Other proteins, such as fibrinogen, are essential to the formation of blood clots. Still others work in conjunction with your body's immune responses.

Plasma also contains ions, lipids, and some dissolved gases, such as oxygen, carbon dioxide, and nitrogen. However, blood solids carry far more dissolved gases than plasma does.

Blood Solids

Blood solids exist essentially suspended within your plasma. In humans, blood solids consist of red blood cells (erythrocytes), white blood cells (leukocytes), and platelets (thrombocytes). All blood solids must be present in order for a human to live, but red blood cells usually take center stage because they're the cells that transport oxygen and carbon dioxide.

Weird Science

At any given moment, there are more than 30 trillion red blood cells circulating in your body, and 2 million of them disintegrate *every second.*

The Red Blood Cells

Red blood cells form in the red marrow of your bones. During formation, an iron-containing protein called hemoglobin also forms. As discussed, it's the hemoglobin in red blood cells that actually transports oxygen and carbon dioxide. During the final formation of red blood cells, their nuclei and organelles dissolve, leaving virtually nothing but hemoglobin. Mature red blood cells are little more than an empty sac that contains hemoglobin.

Essential as hemoglobin is, it doesn't do much for the life expectancy of red blood cells because it cannot repair blood cells when they become damaged, nor can it create new ones. Lacking a nucleus and organelles, red blood cells have no way to repair themselves, and they have a very limited life span, usually no more than 120 or 130 days.

When red blood cells "die," your liver and spleen remove their membranous remains. But, ever efficient organism that your body is), the iron portion of what's left of the hemoglobin molecule gets sent back to your bone marrow (transported by your blood, naturally. All exchanges—nutrients, oxygen, and wastes—are handled by your blood stream, acting as a go-between for your kidneys, liver, lungs, digestive system, and all other organs and cells to be reused in the formation of new red blood cells. Red blood cells appear to have a difficult short life, but compared to white blood cells, they're lolling on the beach and drinking piña coladas.

The White Blood Cells

White blood cells are both the janitors and the SWAT teams of your body. For most white blood cells, life is challenging and short. During times of battle, a white blood cell's life is measured in days, or even hours. But oddly enough, some white blood cells manage to function for years.

Weird Science

White blood cells are considerably larger than red blood cells, and they're also less plentiful. Each cubic millimeter of your blood typically contains about five million red blood cells. That same amount of blood usually contains only 7,000 white blood cells.

Bio Buzz

A **macrophage** is a phagocytic white blood cell that develops from circulating white blood cells and defends body tissues. The word **phagocyte** comes from the Greek words *phagein*, which means "to eat," and *kytos*, which means "vessel."

Five types of white blood cells exist: lymphocytes, monocytes, neutrophils, eosinophils, and basophils. Lymphocytes are further broken down into B cells and T cells, both of which are essential to your immune system. All white blood cells arise from your bone marrow and travel in your circulation highways. But most of their janitorial and defense functions are performed after they've squeezed out of blood capillaries and directly entered into tissues.

Leucocytes are the maids and cops of your body. They're constantly cleaning up, scavenging the dead or damaged cells in your body. They are also always on full alert, ready to respond promptly to any invasion by bacteria and viruses, or any other foreign agent.

Monocytes (mature ones) and neutrophils are the true SWAT teams; they search and destroy anything that shouldn't be in your body. Monocytes follow chemical trails to inflamed tissues, looking for the culprit. They are also phagocytes. Once they spot the problem, they turn into a *macrophage*, which engulfs invaders and debris.

Eosinophils and basophils both play roles in the inflammatory response (a defense against invaders, which is discussed more fully in Chapter 16, "The Immune and Lymphatic Systems"). Eosinophils also function in your immune system, and basophils perform anticlotting functions.

The number of white blood cells in your body will vary, depending upon how active you are, whether you're healthy, or whether you're under siege from invading organisms. The average number of white cells, though, and their constant replacement gives you some idea of just what an inviting place your body is for bacteria, viruses, and protozoans.

The Platelets

Platelets aren't cells, per se. In your bone marrow, some cells develop into "giant" cells, called megakaryocytes. These cells shed fragments of their cytoplasm, which then become enclosed in some of the plasma membrane. These membrane-enclosed fragments become platelets. Platelets are roughly oval shaped, about 2 to 4 micrometers across. They have no nucleus, which gives them a very small life span, usually only about seven days. Although platelets don't live very long, there are a lot of them. A cubic millimeter of your blood will contain approximately a half-million platelets.

Platelets are always circulating in your blood. They are the essential ingredient in the formation of blood clots, which is your body's way to prevent you from bleeding to death if you're cut. When one of your blood vessels rips, tears, or otherwise breaks open, platelets rush to the area, stick together, and form a small plug. The damaged blood vessel then constricts, slowing down the blood coming into the area. Then the platelets release a special clotting factor, which interacts with prothrombin, a protein found in plasma. Together, they form an enzyme that reacts with fibrinogen (the protein found in blood), causing it to form long, sticky threads (called fibrin). The fibrin forms a net that traps red blood cells, and the mass of fibrin and red blood cells then hardens into a clot, or a scab.

In this chapter, we've talked about one of the most complex and encompassing systems your body has: your circulation. With the heart keeping everything moving, blood is the major transport system of any organism's body. In the next chapter, we'll be discussing your immune system. Although your immune system works within your circulatory system, it's so important that it gets its own chapter.

The Least You Need to Know

➤ All animals, from worms to insects, to humans, have a circulatory system consisting of a muscular pumping organ (a heart), blood, and blood vessels.

➤ The human heart is divided into two halves, and each half has an upper chamber (atrium) and a lower chamber (ventricle). This division of the heart is the basis of two cardiovascular circuits: the pulmonary and systemic circuits.

➤ In all vertebrates, the heart pumps blood into arteries, which branch into arterioles, which further branch into capillaries. Capillaries merge into venules, which merge into larger veins, which return the blood to the heart.

➤ Blood is a liquid connective tissue, which is the transport vehicle for carrying nutrients, oxygen, and wastes to and from your cells.

Chapter 16

The Immune and Lymphatic Systems

In This Chapter

➤ Organization makes the human go round

➤ Cells of the immune system

➤ Your lymphatic system

➤ Your immune system in action

➤ Immunity, drugs, and disorders

Your body is a tempting place to live; it attracts all sorts of things to it, including some undesirable tenants. Without some method of screening your "tenants," your survival would be short-lived indeed.

Your body is the equivalent of a fortress; not much gets in because there aren't many ways to get in. But when something does manage to break down the doors, your body is ready and waiting, and your immune system springs into action. In this chapter, we'll be talking about the different defense systems your body has that contribute toward your long life.

Always, Always, Always: Organization!

The Sound of Music isn't the only place where "the hills are alive." Hills, lakes, mountains, plants, air, and even your own body are alive and brimming with microbial organisms. A lot of these microbes are harmless; you and other vertebrates co-evolved with most of them, and you continue to coexist peacefully. But many of them are pathogenic. These are the ones you need to worry about.

When your nose runs, your eyes water, and you sneeze nonstop, you have direct evidence that your body is under attack from a virus, one usually associated with the common cold. It's hard to ignore this kind of attack. But what you may not realize is that every day your body is under attack.

On a daily basis, bacteria, viruses, protozoans, and other foreign bodies make concerted attacks on the organism you think of as "you." Your body provides the perfect environment for these microscopic organisms: it's enclosed, warm, and moist, and it offers a steady supply of food (your cells). But thanks to your immune system, these organisms rarely get to feast upon you.

Like every system in your body, your *immune* system is highly organized. It includes complex chemical and physical barriers, and coordinated cellular responses that work in conjunction with your lymphatic and circulatory systems.

Before discussing how your immune system actually works, it might be helpful to briefly review pathogens themselves. Just what exactly is your immune system responding to?

Bio Buzz

The word **immune** means to be exempt from or protected against something disagreeable or harmful.

Try It Yourself

Over the centuries, science has advanced because people thought and pondered things of interest. Many questions are not immediately answerable, and some may never be. Put on your thinking cap and see if you can advance the scientific cause.

Currently, scientists are attempting to treat allergies by injecting large amounts of the offending allergen into the affected person's blood. What would be the rationale behind this treatment? Do you think it would ultimately work? Could any other body systems be called upon to fight allergies?

Repeated exposure to antibiotics has caused virulent strains of pathogens to develop. Assuming that you can't control the prevalent exposure to these drugs, as a scientist, how would you go about getting the human body to develop better fighting techniques? Do you think there is a way to stimulate the human immune system to fight off invaders, such as AIDS?

Pathogens

A wide variety of organisms function as pathogens to cause human disease, including bacteria, viruses, protozoa, fungi, and worms and other parasites. Let's start with bacteria. In the minds of most people, all bacteria are pathogenic, or disease-causing, but that's not accurate. In fact, only a few types of bacteria cause harm. But the few types that do cause some of the nastiest diseases known. Pathogenic bacteria attack and destroy cells directly. Either they rob your cells of all nutrients, killing them by starvation, or they release toxins (poisonous waste products) that injure or kill the surrounding cells—and the host.

Protozoa carry a number of diseases, with African sleeping sickness and malaria among them. These microscopic animals generally live in the bloodstream, releasing poisons and damaging cells, and sometimes causing death.

Viruses are microscopic infectious particles composed of a core of nucleic acid (either DNA or RNA) surrounded by a protein coat. As you may remember from Chapter 6, "Watch Out! Viruses and Bacteria," viruses enter living cells and take over the cell's machinery, causing the host or infected cell to die. Viruses can produce immediate infections or latent infections. Latent viral infections can live in your body for years, doing no damage. But if something triggers them, such as radiation, they become active, frequently destroying the host. AIDS is a virus, as are most forms of the common cold.

Bio Buzz

A **nonspecific defense mechanism** is a physical, chemical, or biological barrier that serves as a shield against a wide range of pathogens.

Scientists use the word *parasite* to describe many types of organisms that can survive only by living off of another living creature. Tapeworms, which live in intestinal tracts, are one type of parasite. Para-sites usually don't kill their host because that would be self-defeating.

The First Line of Defense: Nonspecific Defenses

The first line of defense your body has against invaders is simple: It tries to keep them out, which is much easier than fighting them once they've taken hold. To this end, the body has several *nonspecific defense mechanisms*.

The largest, most obvious, and best nonspecific defense your body has is your skin. Most pathogens can enter your body only where your skin is broken or in places where your skin is exceptionally thin, such as the mucous membranes of your mouth, nose, or genitals.

Your skin is not only a barrier, but it's an integral part of your immune system. Sebaceous glands located in your skin produce an oily substance known as sebum.

Weird Science

The healing process after a burn offers the best example of the enormous importance of your skin. The first order of business when it comes to helping a person with a serious burn involves growing or grafting new skin onto the burned area as soon as possible. Without skin, you're vulnerable to every pathogen that comes along. Even the most sterile hospital environment can't keep pathogens out the way your skin can.

Think About It!

The swelling and redness that your body produces in response to an pathogenic attack often produces a fever as well. You can take over-the-counter medication (such as aspirin or ibuprofen) to lower your fever, but you may want to think twice before doing so. A fever of 100° Fahrenheit is frequently too hot for most pathogens to function. Given time, your body probably will handily eliminate those pathogens.

Sebum contains several compounds that inhibit the growth of fungi and bacteria. Any rupture in your skin creates an obvious site of entry. But again, your body is well equipped to handle this overt entry of any pathogen.

At places where your skin is thin or where there's an actual entry, your body is waiting with other nonspecific defense mechanisms. Respiratory passages, such as your nose, for instance, are coated with mucous that traps most particles, including pathogens, and then uses ciliated cells to push them into the alimentary canal for elimination. Bacteria that enter through your mouth get pushed into your digestive system, where they are destroyed by digestive enzymes (Remember the hydrochloric acid you read about in Chapter 13, "The Digestive System"? It does more than just break down food.) As an extra defense, the enzyme lysozyme meets bacteria entering through your mouth, urogenital area, and other openings. Lysozyme breaks down the cell walls of bacteria, thereby destroying them. Your eyes are also an entry site for bacteria, but thanks to the presence of lysozyme in your tears, your eyes generally stay free from infection. The inflammatory response is another line of nonspecific defense that your body mounts against any invaders. The inflammatory response is named for obvious reasons: The skin turns flaming red. When your skin is broken, bacteria enter the wound almost immediately. Growing and spreading rapidly, these pathogens begin releasing toxins almost immediately, which kill the surrounding cells.

If bacteria manage to make it through your skin, mast cells (a particular type of immune cell) at the surface of your skin release a substance called histamine. Histamine causes blood vessels near the wound to expand, which causes them to leak fluid. As this fluid enters the wound site, the area begins to swell and turn red.

Histamine and other compounds released at the wound site are also capable of attracting and directing the actions of white blood cells. As the white blood cells come dashing to the rescue, the battle is on. Phagocytes swarm into the area, and monocytes arrive and turn into macrophages. (Remember these guys from your circulatory system?)

As the battle rages, fluid filled with dead bacteria and white cells accumulates at the wound site. The increased cellular activity in the area may even raise the temperature several degrees, causing the area to become reddened, hot, and tender to the touch. Gradually, the bacterial infection is brought under control, and the wound begins to heal.

Your body has other nonspecific tricks up its sleeve. The production of interferon is another nonspecific immune defense available to your body. If a viral infection is present in your body, your cells respond by producing a protein called interferon. Interferon triggers the production of an enzyme that enables a cell to recognize a virus—any type of virus—as a foreign agent. Interferon then prevents the virus from reproducing.

The Second Line of Defense: Specific Defenses

Your nonspecific defenses are just the surface covering—literally. They are your first line of defense against invaders getting into your body. But lurking beneath this nonspecific defense system is an enormous "military" organization, capable of battling very specific invaders.

Your specific defense system recognizes and goes after specific invaders. Specific defenses can function because of the *antigens* found on pathogens. Yes, the same "markers" on your own blood solids are also found on the surface of bacteria, viruses, and other pathogens. *Any* substance—an unfamiliar blood type, lipid, carbohydrate, bacteria, and so on—that can stimulate a specific defense of your immune system is considered to be an antigen.

Your specific defense system targets pathogens in two ways: the humoral immune response and the cell-mediated response. First, the humoral immune response produces a series of soluble, Y-shaped proteins called *antibodies* that circulate throughout your bloodstream. These antibodies bind to sites on any foreign molecules they find. Essentially, antibodies attach themselves to pathogens, waving and yelling at the rest of your immune system, "Yoo-hoo! Over here! Get it!" (Those pathogens can run, but they can't hide.)

Weird Science

Scientists are currently working with DNA-splicing techniques and genetic engineering methods to reproduce human interferon in a laboratory setting. The idea is to deliberately set off an alarm and mobilize cells, thus stimulating the body to fight on its own. So far, this technique has proved effective only in fighting some rare forms of cancer, but scientists are hoping that interferon will prove useful in a broader sense.

Bio Buzz

The word **antigen** is actually an abbreviation of two words: *antibody generator.* Almost any large molecule can serve as an antibody generator.

The humoral immune response is particularly effective against any bacteria or viruses that are in bodily fluids, as opposed to more solid tissues. The term *humoral immune response* gets its name because the antibodies are moving in the *humors* of the body, an old word for the blood and lymph fluids of your body.

Secondly, the cell-mediated immune response organizes a series of cells that will act directly against any foreign cells or tissues that are already infected. This particular defense doesn't discriminate between invaders, but it attacks any foreign substance. This particular immune response is responsible for the rejection of organ transplants. The cell-mediated immune response system doesn't care whether you need that new heart; it's "foreign," and it gets marked as such.

Working together, your humoral and cell-mediated immune responses form the three main functions of your general immune system. These functions include:

➤ **Recognition**—Your immune system can recognize an invader as foreign, or not belonging in your body. Despite the huge diversity of pathogens, your immune system can mount a defensive operation against any foreign organism.

➤ **Reaction**—Once the invader is recognized, your immune system springs into immediate action. Humoral and cell-mediated reactions are organized, frequently involving an attack by specific cells.

➤ **Disposal**—Your immune system doesn't just recognize and react; it destroys. When its actions are effective, the foreign invader is destroyed and completely disposed of.

Cells of the Immune System

Your immune system is responsible for patrolling your entire body. Consequently, its cells aren't confined to just one area, the way your digestive cells are, for example. In that respect, the organization of your

Bio Buzz

An **antibody** is a Y-shaped protein that is capable of binding to an antigen, thereby marking it for destruction by other aspects of your immune system.

Think About It!

If you have an allergy, you have an overindustrious immune system. An allergy is an immune system response that has gone awry. Your body is reacting to a normally harmless antigen (say, pollen, which under normal circumstances is not known for its killer qualities) and is marking it for destruction. In the case of allergies, it's frequently a good idea to disable your histamines (because these are the cells leading the defensive attack) with antihistamines, drugs that can be obtained from your doctor or sometimes over-the-counter.

immune system isn't quite as "tidy" as some of your other systems. Like undercover cops, your immune cells roam everywhere.

Because immune cells have to be able to get anywhere at any time, it's not very surprising that the principal cells of your immune system are lymphocytes, a class of your blood cells. What better way to get around your body? Your circulatory system is the perfect highway.

Like all other blood cells, lymphocytes arise from stem cells in your bone marrow. Under the influence and control of hormones, your stem cells grow, divide, and differentiate into specialized blood cells, of which lymphocytes are merely one type.

Within your immune system, there are two principal kinds of lymphocytes: B-cells and T-cells. Both types of lymphocytes originate in your bone marrow, but it's where they differentiate that distinguishes them. B-cells mature and differentiate in your bone marrow, but T-cells mature only after they also have passed through the thymus gland. Both cells are nearly identical in appearance, but they serve completely different immune functions.

B-cells are responsible for your body's production of antibodies, the molecules that bind to antigens. B-cells are the core of the humoral immune response. T-cells, on the other hand, are responsible for the cell-mediated immune response. The T-cells directly attack the antigens, and the B-cells produce chemicals that render the antigens harmless.

Scientists originally thought that your B- and T-cells worked separately, with no communication or joint effort. But they now know that the functions of both cell types are coordinated. Although they function somewhat separately in what is called the *primary immune response*, T-cells regulate the actions of the B-cells in the *secondary immune response*.

Because they are blood cells, lymphocytes are found throughout your body but are especially present in the lymphatic system, thymus, spleen, tonsils, adenoids, bone marrow, and digestive system.

The third type of cell found in your immune system is the macrophage, or "big eater." Macrophages can engulf and digest all kinds of pathogens and foreign objects. We've already discussed macrophages in Chapter 15, "The Circulatory System."

Bio Buzz

The team effort of your B-cells producing antibodies and your T-cells attacking directly is referred to as the **primary immune response**. The **secondary immune response** is your body's response to any subsequent invasion and infection by the same pathogen.

Your Lymphatic System

Another way your body fights diseases is through the lymphatic system, which is an integral part of your circulatory system. The lymphatic system is an intricate network that works somewhat like a separate highway system specifically set up to handle detours.

By detouring potentially disease-causing organisms off to the side, the lymphatic system can prevent the spread of pathogens throughout your body. The lymphatic system picks up disease agents from the interstitial fluid of your cells and runs it past the cells of your immune system. Then it returns the cleansed fluid back to the circulatory system.

Your lymphatic system operates when some plasma leaks slowly from capillaries into body tissues. The fluid (which at that point is called "lymph") then drains into the lymphatic system and passes through your *lymph nodes*, small organs that filter pathogens from lymph. This drain-and-circulate pattern ensures that at least a few bacteria or virus particles will make contact with a lymphocyte. If a pathogen hasn't directly encountered a lymphocyte in your bloodstream, it will when the normal capillary leaking sends them through your lymphatic system. It's this initial contact with a lymphocyte (B-cell, T-cell, or macrophage) that raises the alarm and sets off the immune response.

When your body is fighting off an infection, your lymph nodes may become swollen. That's because your lymphocytes are detouring so many pathogens through the system that it's having a hard time keeping up.

Bio Buzz

Lymph nodes are small organs that filter lymph. They primarily are located in your groin, neck, and armpit areas.

Your Immune System in Action

When a pathogen manages to enter your body, it doesn't take long for the normal action of your circulatory and lymphatic systems to bring it into contact with one of your immune cells. Here's how.

B-cells Sound the Alarm

Sometimes, B-cells encounter the pathogen with an antigen on its surface. Other times, they're actually presented with the pathogen/antigen by a macrophage that has already engaged the invader in battle.

Weird Science

The antibodies of B-cells are so finely tuned to their particular antigen that they can distinguish between molecules that differ by only a few atoms.

What your immune system does next depends on the type of B-cell involved in the encounter.

Every immature B-cell carries a single type of antibody on its cell surface. One might carry an antibody that responds to the measles virus; another might carry an antibody that responds to the carbohydrates on the cell membrane of a trypanosome (the parasite that causes African sleeping sickness).

This might seem like a hit-or-miss approach to fighting invaders; what if the trypanosome never encounters the "correct" B-cell? But your population of B-cells is huge and diverse. The sheer number of B-cells ensures that an invader will encounter a B-cell that recognizes it and will sound the alarm, while the diversity ensures that your immune system can respond to virtually any antibody/pathogen it might encounter.

Suppose that, just by chance, a B-cell encounters a antigen that matches to its particular antibody. The antibody binds to the antigen and starts a chain of events in the B-cell, which in turn sets your immune response in motion.

Let's assume that a B-cell has encountered an antigen that matches its antibody—in this case, let's say it's a protein. The B-cell immediately digests the antigen (how's that for a "take no prisoners" stance?) into smaller peptides and then displays these peptides on the outer surface of its cell membrane.

If these peptides are recognized by a *helper T-cell* (and again, this is by chance; the correct helper T-cell just happens to be passing by), the T-cell binds to the B-cell and releases interleukins, which will stimulate the B-cell to reproduce into a clustered clone of cells, each of which will produce antibodies against the antigen.

A lot of these new B-cells develop into specialized plasma cells ready to synthesize and release antibodies targeted at the antigen that activated the original B-cell.

Other new B-cells reproduce into clones of memory cells. Your body stores memory cells in the spleen, where they wait, ready to spring into action if they

Bio Buzz

A **helper T-cell** is a type of T-cell that can produce interleukins and stimulate B-cells to reproduce clone cells. They help the B-cells produce more antibodies to fight off the invader, hence the name.

Think About It!

You shouldn't be surprised that it takes so long for your body to fight off a cold virus infection. Although your immune response is quick, your body won't see the actual results for between 5 and 10 days after the initial exposure. That's how long it takes for your body to produce enough antibodies for a full war against the invader.

ever encounter the same antigen again. It's these cells that form the bulk of your secondary immune response. It's also why your secondary immune response is much quicker than your primary immune response, at least when it comes to fighting back. Everything's already set up and waiting. These memory cells are also what provide permanent immunity to some diseases.

T-Cells and the Cell-Mediated Response

The formation of antibodies is clearly a cornerstone of your immune system and its response to invaders. But antibodies aren't the only part of the immune response. T-cells mount a hand-to-hand combat response or, more precisely, a cell-to-cell response. There are three different types of T-cells, each of which performs a different function in the cell-mediated response.

➤ **Helper T-cells**, which stimulate B-cells to reproduce clones and more antibodies

➤ **Cytoxic, or "killer," T-cells**, which directly attack the antigen cell and destroy it by disrupting the function of its cell membrane

➤ **Suppressor T-cells**, which modify and control the rate of response by suppressing the production of antibodies once the infection is under control

Weird Science

So far, scientists have been able to find only two ways to combat your own cell–mediated immune response, when it comes to the rejection of transplanted organs. (It's your cell-mediated response that causes the rejection, since transplanted organs contain markers that cause killer T-cells to spring into action and attack.) First is to use only compatible donors (so that the tissue isn't marked as "foreign"). Second is to suppress the function of your entire immune system through drugs such as cyclosporin, which creates its own set of problems. Any drug that suppresses your entire immune system might solve the problem of organ rejection (in the short run) but leaves you vulnerable to *everything* that invades your body, even seemingly harmless things.

The cell-mediated response of T-cells is vital to your ability to fight off viral infections. Circulating antibodies can trap free-floating viral particles, but they can't do anything about viruses that are already in your cell tissues. If you remember, though, active viruses take over their host's manufacturing system, and this usually causes the host cell to display the virus's proteins on its surface.

Killer T-cells are able to detect these displays, go into the infected cells, and kill them before the virus has a chance to replicate and infect other cells.

Immunity, Drugs, and Disorders

If you're resistant to a particular pathogen, meaning that exposure to it doesn't produce infection or an immune response, you're said to be immune to it. There are two ways to be immune: You can be born with a *natural immunity* to a specific pathogen, or you can *acquire an immunity* by exposure to the substance.

Long ago, natural immunities were illustrative of the phrase "survival of the fittest." In ancient times, children who were born with natural immunities had a better chance of surviving the diseases of early childhood. This still holds true; a natural immunity is a boon.

One way you acquire an immunity is through exposure to the disease itself. You can do this directly (although with some diseases you run the risk of being killed in the process), or you can be vaccinated. With a vaccination, an individual is exposed (usually by means of an injection, although sometimes orally) to a killed or weakened form of the pathogen. This exposure causes your own immune system to create antibodies and memory cells that keep you free from infection if subsequent exposures occur.

It's hard to develop a vaccination or any kind of immunity to the common cold because viruses mutate rapidly. The cold virus to which your body develops an immunity may never show up again because three days after you were exposed, it mutated into something else. You've probably got an acquired immunity to every cold virus you've ever had, but unfortunately (for all of us), those viruses no longer exist.

Bio Buzz

A **natural immunity** is an immunity that is present at birth. An **acquired immunity** is developed after birth.

Vaccines: The Wonder Drugs

Vaccines against disease have become one of the most effective ways of fighting disease because no drugs work effectively on viruses. Drugs called *antibiotics*, however, are very effective against bacteria.

As recently as the 1940s, nearly one-fourth of all the deaths that occurred annually in the United States

Bio Buzz

An **antibiotic** is a chemical that is produced by microorganisms and is capable of inhibiting the growth of many types of bacteria. Antibiotics are the normal metabolic byproduct of certain microorganisms, including fungi.

were the result of pneumonia, tuberculosis, or influenza. Other killers were smallpox, scarlet fever, whooping cough, and diphtheria. Other infections killed countless women during the course of childbirth. Today, most of these same diseases and infections are treated with antibiotics. In fact, scientists have identified more than 2,500 naturally occurring antibiotics, such as penicillin, tetracycline, and streptomycin.

Antibiotics work by interfering with the gene expression and other functions of pathogens. For instance, streptomycin blocks protein synthesis, and penicillin disrupts the bonds necessary to hold molecules together in the cell walls of eubacteria. One type of penicillin causes the cell walls of bacterial pathogens to elongate until it ruptures, sort of stretching them to death.

Antibiotics are truly wonder drugs, but they also have a down side and come with side effects. The direct problem with an antibiotic is that while it kills undesirable bacterial invaders, it also kills all the other beneficial bacteria in your body, sometimes leaving you vulnerable to other infections.

The less direct but more global problem with antibiotics is that they cause bacterial resistance. If even a single cell of invader bacteria survives an attack by your immune system and antibiotics, it most likely did so by mutating. In other words, any future attacks by that antibiotic will be pointless because the invader bacteria can circumvent it through mutation.

When the Immune Response Fails

Your immune system doesn't always function with total precision or success. Obviously, if it did, there would be no sickness among humans. Sometimes, though, the human immune system just goes haywire. Allergies are one such example of an immune system gone haywire, as are autoimmune disorders.

Weird Science

In 1991, the World Health Organization estimated that more than 1.5 million people were infected with HIV. In 2,000, officials revised their estimates and now feel that the number is closer to 10 million. There is currently no vaccine for the HIV, nor is there a cure for those already infected.

When you have an allergy, your body mounts an immune system response to a normally harmless substance. Dust, pollen, perfumes, cosmetics, drugs, and some insect venom are just a few of the things that can trigger an allergic reaction in your body. Some of these reactions occur minutes after exposure, while others are delayed. Either way, allergies can cause tissue damage.

On rare occasions, the inflammatory response to an allergen is so explosive that anaphylactic shock—a life-threatening reaction—occurs. Individuals who are allergic to bee stings are subject to this type of reaction. Within minutes of a single sting, the air passages to their lungs undergo massive constriction, fluid escapes rapidly from their capillaries, blood pressure plummets, and circulatory collapse occurs.

Sometimes the cell-mediated immune response is weakened, and the body becomes highly vulnerable to all sorts of infections that otherwise would not be life-threatening. This is what happens in a person with AIDS. AIDS is caused by the human immunodeficiency virus, or HIV. AIDS is really a constellation of disorders that follows once someone is infected with the HIV, which compromises the immune system by attacking helper T-cells and macrophages. Sometimes the virus attacks the nervous system directly, causing loss of motor functions and mental impairment.

Your immune system is one of the most remarkable systems in your body. Try as they might, with research, drugs, and vaccines, scientists have not been able to duplicate even a fraction of what your body does on a daily basis. They probably never will.

We've discussed an interesting system in this chapter: your immune system. Much of the medical research being conducted these days deals with your immune system and how it functions, particularly because this is the human system that concerns disease most directly. We've talked about different defense systems your body employs to ward off invaders, and different ways it attacks any invaders that get in. In the next chapter, we'll be talking about your reproductive system, which means that we'll be talking about sex!

The Least You Need to Know

➤ The major cells of your immune system are lymphocytes. Lymphocytes are produced in your bone marrow and are a type of blood cell found throughout your body. When a pathogen encounters a lymphocyte, it sets your immune response in action.

➤ Nonspecific defense, such as your skin, keep most pathogens from ever getting into your body. Specific defenses attack pathogens that make it into your body.

➤ Every pathogen carries a specific antigen. An antigen is any large molecule that lymphocytes perceive as foreign. Antigens cause lymphocytes to produce antibodies, which battle the pathogen.

➤ After the initial exposure to a pathogen, antibodies remain in your lymphatic system. If a subsequent exposure occurs, your immune system will spring into action and immediately destroy the pathogen. This phenomenon is called immunity.

Chapter 17

Endocrine and Reproductive Systems

> ## In This Chapter
>
> ➤ Organization: the beginning and the end
>
> ➤ Animals vs. animals
>
> ➤ Your endocrine system
>
> ➤ Human reproductive systems
>
> ➤ Fertilization and development

It's one thing to talk about evolution and the progression of life through the ages. But it's another to consider how it has come about. Evolutionary progress needs more than just the passage of time. It needs reproduction, generations and generations of it.

Every animal on the planet has its own specialized reproductive system, or method for passing on genetic information to future generations. Reproductive systems are responsible for the development of offspring in the image of the reproducing parents. That's what we'll talk about in this chapter.

Always, Always, Always: Organization

The ability to reproduce is the cornerstone of all life. The first cells may have arisen spontaneously, but without the capacity to reproduce, life would have ended right there. Within the animal kingdom, there are two methods of reproduction: asexual and sexual.

In *asexual reproduction*, the offspring arise from a single parent and inherit the genetic traits of that parent only. Asexual offspring are clones, or identical copies of their parents. Asexual reproduction might be "easier," but it doesn't result in variety within a species. Other than the rare mutation, asexual reproduction tends to produce generations of individuals with few variations. Everything from aspens to coral polyps and flatworms engage in asexual reproduction.

This lack of variety is the main pitfall of asexual reproduction. With every individual of a species virtually the same, any sudden environmental change that could kill one can also kill them all. Asexual reproduction works best when the organism is highly adapted to a limited set of environmental conditions. Variation isn't necessary and wouldn't provide the organism with any benefits.

Among animals, sexual reproduction is the dominant form of reproduction. *Sexual reproduction* produces offspring through meiosis, gamete formation, and fertilization. Sexual reproduction requires a species to construct and maintain (at least in an evolutionary sense) specialized reproductive structures, hormonal controls, and complex forms of behavior that are attuned to the environment, potential mates, or both. For a species to engage in sexual reproduction, it must maintain two distinct sexes. Consequently, sexual reproduction is biologically expensive.

Although expensive, this separation of the sexes affords some real advantages, with variety being the main one. Because there is a broad mix of genes, offspring reproduced in this way show a variety of traits, which improves the chances that at least a few of them will survive sudden changes in the environment.

Meiosis and gamete formation occur in two separate, prospective parents. At fertilization, the gamete from one fuses with the gamete of the other, thereby creating a zygote, which is the first cell of the new individual. You have little in common with a frog, and a giraffe has little in common with a polar bear, yet all sexual reproductive systems follow the same basic patterns of organization: gamete formation through meiosis, followed by fertilization.

Variety is an enormous advantage to any species. It would have to be, to offset the cost of sexual reproduction. Consider some of the things that sexual separation entails. Cells that can serve as gametes need to be set aside and nurtured. Specialized reproductive organs need to be developed to house and deliver the gametes. Frequently, mating behavior has to be developed for

Weird Science

Asexual reproduction can take any number of forms. All forms produce offspring that arise from a single parent and contain only the genes of that parent.

Bio Buzz

Sexual reproduction produces offspring through meiosis, gamete formation, and fertilization.

fertilization to occur. And controls have to be developed to ensure the correct timing of gamete formation, as well as the sexual readiness of both partners. Even parental behavior has to be developed.

Among most animals, reproductive timing is critical. Gamete formation and fertilization require a huge outlay of energy. To expend this energy and not have your potential partners in sync would be devastating. Across the evolutionary timeline, animal species that sexually reproduce have developed complex systems for maintaining and operating neural and hormonal mechanisms, with each sex geared to the same cues.

Just getting the timing right for fertilization is hard enough. What about the survival of those offspring? Animals that engage in asexual reproduction invest nothing in the survival of their offspring. But assuring the survival of sexually reproduced offspring is costly.

Many invertebrates and bony fishes release thousands of eggs and motile sperm into their environment. These species have invested energy into the production of gametes, lots and lots of them. Given their environment (usually liquid), these species wouldn't do well if they produced only one egg and one sperm each season.

The reproductive systems of most land animals, though, depend upon internal fertilization, where the union of the gametes takes place inside the female's body. Unlike most water-dwelling animals, these animals have invested energy into the development of elaborate reproductive organs, such as a uterus in females (a chamber where the embryo will grow and develop) and penises in males (a specialized organ for delivering gametes and depositing them within a female).

Many animals also set aside energy to nourish the embryo until it can feed itself. Nearly all animal eggs contain some sort of yolk, a protein-rich lipid substance that will sustain the developing embryo. Human eggs contain very little yolk, but human females invest a lot of energy into the development of a placenta and other tissues while pregnant.

Think About It!

Many animal species use seasonal changes, such as the amount of daylight, to coordinate the timing of their reproductive systems. In male and female moose, for instance, fertilization occurs only in late summer and early fall. The rest of the year, the animals are sexually inactive. This coordination within the species ensures that offspring won't be born until the following spring, when food is plentiful.

Weird Science

The amount of yolk found in animal eggs varies. Sea urchin eggs contain very little yolk. Urchins put their energy into producing thousands of eggs, and put very little energy into sustaining those eggs. Bird eggs, on the other hand, are very yolky. Birds put their energy into the production of only a few eggs, and a larger yolk helps ensure that each egg will survive.

Humans and most other animals go through six stages of development.

1. Gamete formation
2. Fertilization
3. Cleavage
4. Gastrulation
5. Organ formation
6. Growth and tissue specialization

The first stage, gamete formation, involves the development of gametes (sperm and eggs) inside the respective bodies of the prospective parents. During the second stage, fertilization, the plasma membrane of the sperm fuses with the plasma membrane of the egg. The egg nucleus and sperm nucleus then fuse, creating a zygote.

In the third stage, cleavage, mitotic cell division divides the zygote into smaller, nucleated cells. Each of these cells is called a blastomere. The cleavage stage of embryo development creates only more cells; it doesn't create more volume. During cleavage, the size of the zygote and the volume of its cytoplasm remains the same—it just gets divided into smaller pieces.

As the cleavage stage draws to a close, the pace of mitotic cell division slows down, and the embryo enters the gastrulation phase of development. During gastrulation, the embryo undergoes major cellular reorganization, with all the cells rearranging themselves into one of three types of primary tissue. Out of these three primary tissues, or germ layers, will come all the organs and tissues of the adult animal. The three types of tissue are these:

Weird Science

The mesoderm tissue layer is a real evolutionary development. It originated millions of years ago and is responsible for the development of nearly all large, complex animal species. Without the development of this complex tissue layer, we'd probably all still be sac-gut marine organisms.

1. **Endoderm**—This produces the inner lining of the gut and the organs derived from the gut. Endoderm is the innermost tissue layer.

2. **Mesoderm**—This produces "middle" tissues such as the skeleton, muscles, circulatory, and reproductive systems. Mesoderm is the intermediate tissue layer.

3. **Ectoderm**—This produces the nervous system and the outer layer of the integumentary system. Ectoderm is the outermost tissue layer.

After these three primary tissue layers have formed, the embryo moves on to the next stage of development: organ formation. The primary tissue layers give rise to subpopulations of cells. During organ formation, these subpopulations develop into unique structures with special functions. The descendants of these cells give rise to the different types of tissues and organs found in the adult animal's body.

The final stage of animal development is the growth and tissue specialization stage. This is the stage in which the organs and tissues gradually increase in size and assume their specialized functions. The growth and tissue development stage continues into adulthood.

Although each of the six stages of animal development is separate and unique, the development of the individual cannot (or will not) proceed correctly unless each stage is successfully completed before going onto the next stage.

The stages of animal development are interesting from an evolutionary standpoint. Research has shown that the first stages of development are open to evolutionary change, but that the early development of organs is not, which is probably why we see only a handful of body plans among animals. There have always been physical constraints (such as surface-to-volume ratios) and architectural constraints (such as those imposed by the body's axis), but it now appears that there are also master genetic constraints. Once the interactions needed to produce organ structures have started, it's difficult to start over again.

Your Endocrine System

As complex as reproductive systems are, another component to reproduction can be even more complex because it deals with the hormones, or chemical messages that control your reproductive system. The endocrine system is responsible for the production of hormones, and in humans, the reproductive system is linked to the endocrine system. Your endocrine system consists of glands that transmit chemical messages throughout your body. These chemical messages are referred to as *hormones*. Hormones control much of the sexual reproductive cycle in all animals, not just humans.

The Wonderful World of Hormones

Different hormones are produced by different glands. The glands secrete tiny amounts of the hormone into your bloodstream, which carries these chemicals to various parts of the body. Even though hormones circulate throughout your body, they affect only specific cells, called target cells, that have receptors for that particular hormone. Unless a cell has a receptor and is actually a target cell, no amount of any hormone will influence its activity.

Bio Buzz

Hormones are substances that are produced in organs of the endocrine system in one part of the body that influence the activity of cells in another part of the body.

Hormones differ in their chemical structure, but there are two main types in your body: protein hormones and lipid hormones. Most of the hormones produced by your endocrine system are protein hormones. Protein hormones attach to receptors on target cells, where the hormone then activates an enzyme in the target cell's membrane. The enzyme then causes the bodily changes associated with the hormone.

Lipid hormones are also called steroids. Steroids diffuse through the target cell's membrane and enter the cytoplasm. Once in the cytoplasm, they bind to protein receptor molecules, forming a hormone-receptor complex that enters the cell's nucleus, where it stimulates genes to produce messenger RNA. In turn, the messenger RNA forms proteins that bring about the changes associated with the hormone.

The Glands That Secrete the Hormones

Endocrine glands operate through feedback mechanisms. A feedback mechanism is a system in which the end product of a series of steps controls the first step in the series. If the end product promotes the first step in the series, it's called a positive feedback mechanism. If the end product inhibits the first step in the series, it's called a negative feedback mechanism.

Most endocrine glands operate through a negative feedback mechanism. Think about how a thermostat controls a furnace and the heat in your house. If the temperature drops below a specified level, the thermostat signals for the furnace to turn on and warm up the air. The end product of this series of steps—hot air—is detected by the thermostat, which then proceeds to turn off the furnace. That's negative feedback at its simplest.

Your endocrine system works something like a thermostat. If the amount of a particular hormone drops in your blood, your endocrine system senses it and secretes more into your bloodstream. The increased hormone causes the activity of target cells to increase, which in turn will cause your endocrine to cut back on hormone secretion.

Your body has two types of glands: *exocrine* and *endocrine*.

Your sweat glands are exocrine glands; they release all their secretions through ducts. Endocrine glands are found all over your body. The human body has approximately 10 endocrine glands: the pineal, hypothalmus, and pituitary glands are found in the head. The parathyroid and thyroid glands are found in the neck. The thymus, adrenal, Islets of Langerhans, ovaries, and testes are found in the torso. Each of your endocrine glands performs a different function.

Bio Buzz

Exocrine glands release their secretions through ducts. **Endocrine glands** release their secretions directly into the bloodstream.

Try It Yourself

Over the centuries, science has advanced because people thought and pondered things of interest. Many questions are not immediately answerable, and some may never be. Put on your thinking cap and see if you can advance the scientific cause.

Temperature affects more than just human sperm. In alligators and sea turtles, the temperature of the ground in which eggs are laid will affect the sex of the offspring. A temperature difference of only a few degrees will cause an egg clutch to produce all males, or conversely, all females. Why do you think this would happen? What would the temperature be affecting: the sperm, the egg, or the zygote itself? Do you think temperature therapy could be used to affect the gender of human embryos?

Your thyroid gland secretes the hormone thyroxine, which plays a vital role in regulating protein synthesis and ATP production. If your thyroid gland produces too much thyroxine, hyperthyroidism develops. People with hyperthyroidism are physically overactive and have high blood pressure and higher body temperatures; their hearts also beat faster than normal. If the thyroid gland produces too little thyroxine, hypothyroidism develops. People with hypothyroidism tend to be overweight and to tire easily.

Parathyroid glands regulate the amount of calcium and phosphate ions in the blood. When blood levels drop too low, your parathyroid glands secrete parathyroid hormone, which causes target cells in your bones to release calcium and phosphate.

Your adrenal glands, as their name indicates, are responsible for adrenaline, the stuff that causes your bronchial tubes to enlarge, your pupils to dilate, your heart to beat faster, and your blood pressure to rise.

The thymus gland is larger in children than in adults. It produces the hormone thymosin, which stimulates the development of infection-fighting antibodies and boosts a child's immune system. As an individual matures, the thymus decreases in size, having fulfilled its primary function.

The Islets of Langerhans are actually specialized cells within your pancreas. They secrete the hormones insulin and glucagon. Insulin lowers your blood sugar, while glucagon raises it. Disorders of this gland usually take the form of diabetes.

Although gonads are endocrine glands, they are also the gamete-producing organs of your reproductive system. Female gonads are referred to as *ovaries,* while male gonads

Think About It!

Don't forget the salt! Your body needs iodine to function. Insufficient amounts of iodine cause the thyroid gland to swell, a condition known as a goiter. In the past, this was a frequent occurrence in people who lived in areas where the soil contained little iodine. Thyroxine, a hormone produced by your thyroid gland, produces iodine, but without getting some iodine in your diet (either directly or by consuming plants that have absorbed it through the soil), your body doesn't get enough iodine to function correctly. Nowadays, with the addition of iodine to table salt and drinking water, goiters are not common.

Bio Buzz

An **ovary** is a primary reproductive organ responsible for the formation of eggs. A **testis** (plural, *testes*) is a primary reproductive organ, which is responsible for the production of sperm.

are referred to as *testes*. Gonads produce gametes, and they also produce hormones, which is why they are part of both the reproductive and the endocrine systems.

Ovaries secrete hormones called estrogens and progesterone. In females, these hormones cause breasts to develop, hips to widen, and menstruation to begin. Testes secrete a group of hormones called androgens. The most common androgen is testosterone. In males, hormones generated by the testes cause the voice to deepen, the chest to broaden, and more hair to grow on the body.

Together, your pituitary gland and hypothalamus regulate the activity of all your endocrine glands and control body growth.

Human Reproductive Systems

In both men and women, the reproductive system consists of a pair of primary reproductive organs (gonads) and a number of accessory glands and ducts. Taken altogether, the human reproductive system produces gametes, stores and releases sex cells, and nourishes and protects the developing embryo.

The Male Reproductive System

The male reproductive system produces, stores, and releases male gametes, called sperm. Men start producing sperm during puberty and continue to do so for most of their lives. The primary function of sperm is to store and deliver genetic information to the female egg. And its shape is eminently suited for this function.

A sperm cell can be divided into three regions: head, mitochondria, and flagellum. The large head region of a sperm contains DNA. The narrow middle section, located just below the head, contains mitochondria. The mitochondria are the motor; they supply the energy needed to move the sperm to the egg. The flagellum, or "tail," whips about (using energy supplied by the mitochondria) and propels the sperm forward.

Sperm begin life in one of two testes, which are the male gonads. Both testes are contained within the scrotum, a pouch of skin that's an extension of the lower abdominal wall. The scrotum holds the testes outside the abdominal wall rather than inside the body. The temperature inside the scrotum is slightly lower than your body temperature, which promotes the development and survival of sperm.

The testes are composed of tightly coiled tubes called seminiferous tubules. The seminiferous tubules are where special cells divide by meiosis and form into sperm. The newly formed sperm are then released into an elongated sac called the epididymus, for maturation and storage. It takes about 18 hours for sperm to mature and become fully functional.

When a man becomes sexually aroused, contractions in the muscles of the walls of the epididymus push the mature sperm into a pair of thick-walled tubes called the vas deferens. From there, more contractions propel the sperm through ejaculatory ducts and then through the urethra, which are located in the penis, a specialized reproductive organ for delivering sperm into a female. During arousal, its spongy tissue fills with blood and becomes stiff and erect.

As the sperm travels to the urethra, it mixes with glandular secretions to form semen. Suspended in semen, the sperm is both protected and nourished. A primary component of semen is fructose, a sugar that the sperm uses for nourishment. Semen also contains a buffering ingredient that protects the sperm from acidic conditions they might encounter on their way to an egg.

The Female Reproductive System

In females, the ovaries produce and store eggs, the female gametes. An egg is referred to as an ovum. Human females are born with more than 400,000 immature eggs in their ovaries and don't produce any more during their lifetime. Over the course of most women's lifetimes, only about 400 eggs will actually mature.

The female reproductive system contains two ovaries, two Fallopian tubes, a uterus, a cervix, a vagina, and external genitalia. The Fallopian tubes are located between the ovaries and the uterus, and deliver mature eggs to the uterus. Fertilization can occur in either the Fallopian tubes or the uterus.

Weird Science

Sperm need a temperature of approximately 95° Fahrenheit to survive. To keep sperm at this temperature, a specialized control mechanism operates muscles in the scrotum. When the air outside the body is cold, muscle contractions draw the scrotum closer to the body. When the outside temperature is warm, the muscles relax and lower the pouch.

Weird Science

An ovum is just barely visible to the naked eye. It's teeny. Yet an ovum is 75,000 times larger than a single sperm cell.

Think About It!

Reproduction is a natural part of being human, yet many couples who desperately want to have children find themselves experiencing problems conceiving or carrying a pregnancy to full term. To find out more about infertility and solutions to this problem, check out the National Institutes of Health Web site at http://www.nih.gov/.

Bio Buzz

The **menstrual cycle** usually occurs in cycles of 28 days. During this time, an ovum matures and is positioned to meet a sperm in the Fallopian tube. The ovum will travel through the Fallopian tube into the uterus. If fertilization does not occur, the egg will be discharged through the vagina.

The uterus is a muscular saclike organ that will house and protect the embryo if fertilization occurs. The lower entrance of the uterus is called the cervix. The cervix leads into the vagina, a tubelike canal that leads to the outside of the body. The vagina is the canal that accepts the penis during intercourse and through which a fetus will pass during childbirth.

Every month, the female reproductive system goes through a series of chemical and physical changes called the *menstrual cycle*. The menstrual cycle has four phases:

1. The follicular phase
2. Ovulation
3. The luteal phase
4. Menstruation

The follicular phase begins when the hypothalamus stimulates the pituitary gland to produce follicle-stimulating hormone (FSH). FSH travels through the bloodstream and stimulates a group of ovarian cells, called a follicle, to form around the ovum. FSH also promotes the production of another hormone, estrogen. Estrogen releases into the bloodstream, where it stimulates the uterus to produce a thick lining of tissue, mucus, and blood vessels. Estrogen also causes the pituitary gland to produce luteinizing hormone (LH), which causes the ovum to mature. (Bear in mind that this is only the first part of the menstrual cycle. No one ever said that sexual reproduction systems were simple.)

During the next phase, ovulation, the follicle around the egg (which is now mature, thanks to LH) ruptures and releases the ripe egg. Fingerlike projections draw the egg out of the ovary and into the Fallopian tube. The egg remains in the Fallopian tube for about four days, during which time fertilization can occur. If the egg becomes fertilized, the resultant zygote moves through the Fallopian tube and into the uterus.

In the luteal phase of the menstrual cycle, the ruptured follicle that was left on the surface of the ovary develops into a completely new structure, the corpus luteum, which is a glandular structure. Under the influence of the pituitary gland, the corpus luteum sends out steroid hormones, including estrogen and progesterone. These hormones then stimulate the lining of the uterine wall to become even thicker. The rising levels of these hormones causes the body to stop producing FSH.

The last phase of the cycle is menstruation. If the egg was fertilized, it will attach to the uterine lining. If it wasn't fertilized, it won't attach. In the case of an unfertilized egg, the corpus luteum stops producing progesterone, and the uterine lining sloughs off. The lining, along with the egg, is discharged through the cervix and vagina. After menstruation, the cycle begins all over again.

Fertilization and Development

The main function of any reproductive system is to create new individuals. When a male gamete (sperm) successfully fuses with a female gamete (an ovum), the resulting zygote develops into a new human being.

Nine Months and Counting

Human embryos spend approximately 9 months, or 38 weeks, developing in the uterus. By the time fertilization occurs, the new zygote has already progressed through two of the six steps of development: gamete formation and fertilization. Like most other animals that are the product of sexual reproduction, the human embryo also progresses through the remaining four steps: cleavage, gastrulation, organ formation, as well as growth and tissue specialization.

Throughout the pregnancy, the growing embryo is surrounded by four membranes. The first membrane is the chorion, which possesses small finger-like projections called chorionic villi. Waste products from the embryo are exchanged with food and oxygen from the mother through the membranes of the chorionic villi. Where the chorionic villi meet the maternal blood supply is called the *placenta*. The placenta is the life-supporting link between the mother and the developing child.

Weird Science

The time it takes to progress through the six stages of embryonic development varies from animal to animal. In less than three weeks after fertilization, a frog embryo turns into a swimming, algae-eating tadpole. Three weeks after fertilization, a human embryo is already recognizable as a vertebrate; eight weeks after fertilization, the embryo is recognizable as a human being.

Bio Buzz

The **placenta** is a blood-engorged organ that develops in pregnant female mammals. It permits exchanges between the mother and the fetus without intermingling their blood. The new individual is sustained, yet it develops a circulatory system and blood type that is separate from the mother.

The second membrane is the amnion. It is a sac filled with amniotic fluid, which cushions the embryo and keeps it moist. The third membrane is the yolk sac, which provides some nourishment during early development. Finally, the fourth membrane surrounding the developing embryo is the allantois. With the chorion, the allantois grows and lengthens into the umbilical cord. The umbilical cord contains arteries and veins that carry blood between the placenta and the embryo.

In human embryos, the most dramatic developmental changes take place within the first eight weeks. By the end of the fourth week, the embryo is 500 times its original size. Limbs form, fingers and toes emerge, and the circulatory system becomes intricate. For humans, the embryonic period ends at eight weeks. After that, the embryo is referred to as a fetus. For the rest of the pregnancy, cells will continue to divide, organs to form, and tissues to specialize, preparing the fetus for life in the outside world. By the time birth occurs, a healthy fetus will weigh approximately 8 pounds.

Think About It!

Scientists previously thought that the placenta was a barrier between the mother and developing child. A woman could drink all the martinis she wanted or take whatever drugs were prescribed, they thought, and none of this would affect the baby. This thinking came to an abrupt end in the 1950s when thalidomide, a tranquilizer prescribed for nausea in the first trimester, was discovered to cause birth defects. Women who had taken the drug gave birth to children missing arms and legs, or with grossly deformed toes. Likewise, alcohol has been discovered to cause birth defects. If you are pregnant, you should not consume alcohol, nor take any drug without consulting your doctor.

Into the World

Toward the end of pregnancy, the uterus will start having mild contractions. Relaxin, a peptide hormone secreted by the corpus luteum, will cause the cervix to soften and the connective tissue around it to loosen. Relaxin also loosens the connective tissues between the pelvic bones. All these changes signal the start of the birth process. Uterine contractions become stronger and more rhythmic, gradually increasing in intensity. Eventually, the contractions expel the fetus from the uterus, through the vaginal birth canal.

Following birth, the baby's umbilical cord is tied off and then cut. The baby's lungs expand for the first time, and it begins to breathe on its own. The placenta is usually expelled from the uterus within 10 minutes of the baby's birth.

Thanks to sexual reproduction, every baby is unique—even when it's an identical twin. Sexual reproductive systems may be biologically costly, but they produce diversity in all animals.

Reproduction is an intricate, fascinating process that perpetuates and advances every species. In this chapter, we talked about the basics of sexual reproduction, as well as its advantages and drawbacks to any species that reproduces sexually. In the next chapter, we'll be switching gears and talking about something entirely different: the secret life of plants.

> ### The Least You Need to Know
>
> ➤ In animals, sexual reproduction is the dominant method of reproduction.
>
> ➤ Unlike asexual reproduction, in which the offspring is a clone of the parent, sexual reproduction produces enormous diversity among individual offspring.
>
> ➤ Humans have a primary set of reproductive organs, accessory ducts, and glands. Both males and females have gonads, which are the gamete-producing organs. In females, the ovaries produce eggs, or ovum; in males, the testes produce sperm.

Part 4

How Plants Live

An "Attack of the Killer Tomatoes" may sound far-fetched, but read on. Plants are so diverse, adaptive, and far-flung, scientists know they have not yet discovered all the different species on the planet. They also know they have yet to discover all the uses for the plants they do know about. There may very well be an unknown species out there—a species the marines would love to train for combat missions.

In fact, there is no such thing as a typical plant because they're all different. But there are some common traits that all plants share, the biggest one being that they can make their own food. Just give them a little light, and off they go. Most plants also possess similar tissue structures, and all plants can reproduce (not only that, they re-produce in a way that's remarkably similar to animals, including yourself).

Plants are the backbone of our biosphere. Alive in their own right, they also ensure that all the rest of us stay alive. Because of the oxygen plants create, one could even speculate that animals might never have evolved if plants hadn't been there first. You owe your life to a plant, as you'll discover in Part 4.

Chapter 18

The Plant Body

In This Chapter

➤ The "typical" plant

➤ Which plant has what tissue

➤ Meristems: where it all starts

➤ Roots, stems, and leaves: the basics

More than 350,000 different species of plants exist on Earth. Plants are found everywhere, from tropical rainforests, to temperate meadows, to glaciers. They even live in oceans, lakes, and rivers. Out of all plant species, vascular seed-bearing plants have been the most successful in terms of diversity and distribution. Most of this success can be attributed to their structure and physiology.

Plant structure and physiology is a complex subject (each aspect is a field of study unto itself). But like most things, it's not too hard to understand when you take it one step at a time. In this chapter, we'll be tackling different aspects of plant development, one at a time. We'll be talking about the general tissue structure of plants, how they grow, and the basic structures of a plant: roots, stems, and leaves.

The "Typical" Plant

There's no such thing as a typical plant—just glance back through Chapter 9, "Green, Green, Green: Plants" to refresh your memory. Plants are nothing if not diverse. Even so, when we hear the word *plant*, what usually comes to mind is a seed-bearing *vascular plant*. Vascular seed-bearing plants fall into two categories: *gymnosperms* and *angiosperms*.

Angiosperms include such diverse plants as roses, corn, cacti, bamboo, and elm trees. Out of the more than 350,000 known plant species, at least 235,000 of these are angiosperms. Clearly, angiosperms dominate the plant kingdom.

Within the angiosperm group, there are two main classes of flowering plants: *monocots* and *dicots*. Because both are types of vascular flowering plant, monocots and dicots have many similarities. They both have the same tissues, but these tissues are organized in different ways. Let's take a look.

Monocots

With only about 90,000 species, monocots aren't as numerous as dicots. But among these 90,000 species are some of the world's most important food crops, crops that, in the technical sense, are grasses, such as oats, corn, wheat, rye, and rice.

Cotyledons form in seeds as part of the plant embryo. Their function is to store and absorb food for the developing embryo. When the seed germinates, the cotyledons whither away, and the plant's leaves start functioning. Many monocots, particularly the food crops, were domesticated and cultivated by humans. Wheat was one of the first crops to be domesticated, about 11,000 years ago in the Middle East. Because of human intervention, the wheat we eat today bears little resemblance to the original wheat that first developed on Earth.

Dicots

Dicots are the most numerous plants on Earth (other than angiosperms in general). They are found everywhere, and their success dates back a long way—to the dinosaurs, in fact. They also comprise the majority of angiosperms. Lettuce, maples, cotton, carrots, and daisies are all dicots.

As previously mentioned, the primary difference between monocots and dicots is not in their tissue structure, but in the way they are arranged. Dicots also have cotyledons, but instead of one, they have two. They also have flowers, but instead of being arranged in multiples of three, dicot flowers usually come in four or five parts, or multiples thereof. Likewise, all dicots have leaves, but the veins on their leaves are arranged in a netlike pattern.

Bio Buzz

A **vascular plant** is a plant with conducting tissues and well-developed roots, stems, and leaves. A **gymnosperm** is a vascular plant that bears seeds on the exposed surfaces of reproductive structures, such as cone scales. An **angiosperm** is a flowering plant.

Bio Buzz

A **monocot** is a vascular flowering plant with only a single cotyledon (a seed-leaf, or a tiny leaf that develops in the seed itself) in its seed. The flowers of monocots usually come in multiples of three, and the veins on monocot leaves are arranged in a parallel fashion.

A **dicot** is a flowering plant with two cotyledons in its seed. Dicots also have net-veined leaves, and flowers are arranged in fours, fives, or multiples thereof.

Angiosperms in general, and dicots in particular, owe their success to the fact that they evolved at the same time insects were evolving. Besides having the enormous advantage of seeds, they also had the advantage of pollination by insects.

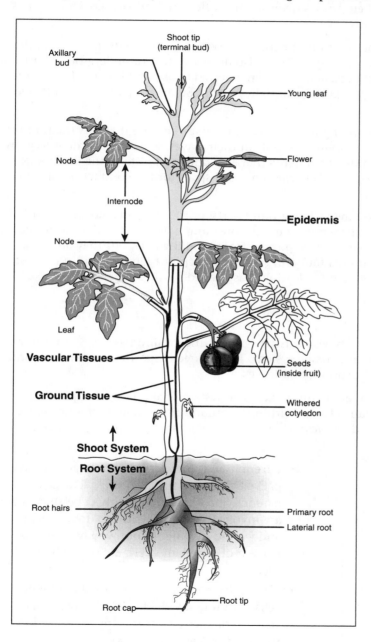

Most flowering plants have body plans similar to the tomato. They have vascular tissues, root systems, and shoot systems comprised of stems and leaves.

235

Tissue, Tissue ... Which Plant's Got the Tissue?

There may be no such thing as a "typical" plant, but there are some common characteristics. In the midst of enormous diversity, most flowering plants have body plans similar to that of a tomato.

Flowering plants have above-ground parts, or shoots, such as stems, leaves, and flowers or other reproductive structures. They also have a system of underground parts, or roots. In plants, roots, shoots, and leaves are referred to as plant organs. The shoots of flowering plants offer support for upright growth. Stems are particularly adapted for this.

Upright growth gives flowering plants a real advantage because it allows them to get more sunlight for the photosynthetic cells in their leaves, which helps them to grow better. (Think of forest mosses and ferns that are confined to life in the shadow of flowering plants.) Stems also provide an enclosed space for the conduction of water, ions, and other nutrients.

All flowering plants also have below-ground parts, or root systems. Roots are specialized plant structures that penetrate the soil, spreading downward and outward. Roots serve to anchor large plants and keep them from toppling over, but they also absorb water, minerals, and nutrients found in soil and transport them upward to the rest of the plant. Roots also store food, releasing it as needed.

Ground Tissues

All plants have three main types of tissue systems: the ground tissue system, the vascular system, and the dermal tissue system. These three tissue systems can be found in every plant part: roots, shoots, and leaves.

The ground tissue system is the most extensive system of any flowering plant. It comprises the bulk of the plant's body. Within the ground tissue system of a flowering plant are three specialized *simple tissues*: parenchyma tissue, collenchyma tissue, and sclerenchyma tissue.

Bio Buzz

In plants, **simple tissues** are composed of only one type of cell.

➤ **Parenchyma tissue** consists of large, loosely packed cells with thin cell walls. The cells of parenchyma tissue are roughly cube-shaped and are involved in photosynthesis, food storage, and wound-healing. In nonwoody plants, 80 percent of the plant's cells are parenchyma tissue.

➤ **Collenchyma tissue** is composed of elongated cells with roughly thickened, flexible walls. The cells of collenchyma tissue are specialized to support the plant as it grows.

➤ **Sclerenchyma tissue** is mostly recognized as a mature tissue. At maturity, sclerenchyma tissue is made up of the walls of dead cells that support and strengthen the plant. Sclerenchyma tissue doesn't stretch, so it's primarily found in the regions of the plant that have stopped growing.

Ground tissues are experts at adapting to meet environmental conditions. All ground tissues are composed of only one type of cell, but those cells can take many forms. Cacti live in environments where water is scarce, so they need to conserve all the water they come in contact with. That's why they have large amounts of parenchyma tissue, which stores water. Herbs exposed to wind need to be flexible, so they possess large amounts of collenchyma tissue that support them. And sclerenchyma tissue can be found where hardness would be an advantage for the plant, such as in the spines of cacti or the shells of nuts.

Vascular Tissues

All parts of a flowering plant contain vascular tissues. Unlike ground tissues, vascular tissues are referred to as *complex tissues*.

Two types of vascular tissue exist: xylem and phloem. Both function to distribute substances throughout the plant. Frequently, the conducting cells of vascular tissue are associated with parenchyma cells and fiber sheaths.

Xylem conducts water and minerals. Mechanically, xylem also helps to support the plant. Xylem is composed of two types of conducting cells: vessel members and tacheids. (For our purposes, you don't need to pay much attention to either of these cells; merely know that two types of cells compose xylem.) Both types of xylem cells have a limited life span: They die when fully developed. By the time they reach full maturity, both types of cells are dead, and their walls are interconnected. Taken altogether, these interconnected walls strengthen the plant and form water-conducting pipelines. Numerous pits in the cell walls allow water to flow in and out of adjoining cells, thus

Think About It!

The next time you wear a linen shirt, tie up a package with some rope, or write a note on a piece of paper, think about where it comes from. Some cells in mature plant tissue (sclerenchyma) are long and thin, or what we refer to as "fibers." The fibers found in cotton are widely used in clothing. Fibers found in the stem of the flax plant are used for linen, and fibers from hemp are turned into rope. Fibers from the papyrus reed can be turned into paper, or even boats.

Bio Buzz

In plants, **complex tissues** are composed of a variety of cell types. Both vascular and dermal tissues are complex tissues.

supplying every plant part with water. Xylem is what keeps plants—and especially trees—standing upright; without it, they'd flop over.

Phloem conducts sugar and other solutes throughout the plant. It works to conduct cells that are referred to as sieve-tube members, and unlike xylem cells, phloem cells are alive at maturity. Sieve-tube members stack end to end, forming a long tube. Compounds conducted through the plant move through regions at the end of each sieve-tube member, called sieve plates.

Sieve-tube members are always found next to specialized companion cells. Botanists believe that these companion cells are responsible for helping to control the movement of sugars through the sieve-tubes.

Bio Buzz

Stomata are structures shaped like holes in the epidermis of plants. By opening and closing, the stomata regulate the passage of gases in and out of the plant.

Bio Buzz

A **meristem** is a localized region of embryonic, self-perpetuating cells. Meristems can be either **apical** or **lateral.** Apical meristems make plants grow tall, while lateral meristems make them grow bigger and rounder.

Like ground tissue, vascular tissue is also capable of modifying itself for adaptive advantage. In trees, xylem forms the wood, providing strength and support while conducting water at the same time. In aquatic plants, xylem isn't needed for support and may be entirely absent in mature plants.

Dermal Tissues

The dermal tissue system in plants provides an outside covering. As in humans, a plant's dermal tissue is composed of an outer layer of cells called the epidermis. In turn, the epidermis is covered with a waxy cuticle. The epidermis protects the plant, helps prevent water loss, and functions in gas exchange through openings called *stomata.*

In plants with mature stems and roots, the epidermis eventually is replaced by cork, another type of dermal tissue. Cork cells usually are packed tightly together, and their cell walls contain a fatty substance called subarin. The dermal tissue of plants is a whiz at adaptation. In dry environments, for instance, the stomata sink so far into the epidermis that water loss is nearly impossible. Although stomata usually are found on the underside of leaves, the stomata are found on the upper side of the leaves in water lilies, which allows gases to exchange with the air; otherwise they plants would "drown."

Meristems: Where It All Starts

Most plants don't grow all the time, even those in the tropics. Even during the growing season, vascular flowering plants don't grow in all parts at the same time.

For instance, if you nailed a sign on a tree trunk—let's say, 1 meter above the ground—it would stay right there, no matter how tall the tree grew. That's because, in plants, growth is confined to localized regions of specialized cells called *meristems*. Meristems can be either *apical*, which produce lengthening growth, or *lateral*, which produce growth that increases the circumference of a plant.

The lengthening, or growth, of all shoots and roots originates in the dome-shaped tip of apical meristems. Overall, this lengthening of the plant, both upward and downward, represents its primary growth. Apical meristems are found in the tips of roots and the tips of shoots.

Besides growing upward into shoots and downward into roots, during the growing season, the stems and roots of most plants also thicken. Lateral meristems are responsible for this increase in circumference. The vascular cambium lateral meristem produces secondary vascular tissues, while the other lateral meristem, the cork cambium, produces a sturdier covering that replaces the epidermis. This thickening of the stems and roots by lateral meristems is also referred to as secondary growth.

Roots, Stems, and Leaves: The Basics

In plants, the tissue systems (ground, vascular, and dermal) arrange themselves into organs, including roots, stems, and leaves. All flowering plants have roots, stems, and leaves. They may look different, but all the parts are there.

Roots

All vascular plants have roots. The plants themselves may look vastly different, and even their roots may look different, but all roots perform the same functions: They anchor the plant, they absorb water and dissolved mineral ions, and they store food made during photosynthesis.

Roots are the first thing to emerge from a seed. The primary root emerges from the seed coat and starts growing downward. But after that, some roots look and act different. In dicots, the primary root increases in diameter as it grows downward. Pretty soon, lateral roots start forming in the root's internal tissues and erupt through the epidermis. The youngest lateral roots are closest to the growing, pushing root tip, while the oldest ones are closest to the base. A plant that has a primary root with lateral branching has a taproot system. Taproot systems obtain water from deep in the ground.

Monocots also have primary roots, but the initial root doesn't live very long because it tends to get lost in the crowd. Instead of a primary root with lateral roots branching off, adventitious roots grow from the plant's stem (at the surface of the soil), and then lateral roots branch off these adventitious roots. In monocots, all the lateral roots are roughly alike in diameter and length. This type of root system is called a fibrous root system. Fibrous root systems obtain water by spreading out, but they don't go as deep as taproot systems.

Think About It!

If you've ever had trouble trans-planting your rhododendron, you can blame its root hairs. They anchor a plant firmly in the ground, spread out everywhere, and grip the soil for dear life. Because root hairs are instrumental in gathering water and nutrients, disrupting them can send a plant into shock. Only a first-time gardener would just yank a plant from the ground when transplanting. Instead, try digging a deep circle around the plant, then gently prying the plant loose.

The structure of roots is also similar, regardless of the type of vascular plant. The apical meristems of roots produce cells that will differentiate into ground tissue, vascular tissue, and dermal tissue. Each of these three tissue systems then differentiates into cells that form the mature root of the plant.

The root tip is organized into three sections:

1. **Apical meristem**—This is the region of cell division, where all cells of the root system originate.

2. **Region of elongation**—Cells from the apical meristem elongate here, and the root increases in length. This region is directly behind the apical meristem and in front of the region of maturation.

3. **Region of maturation**—Here the root cells mature and differentiate into tissue systems.

The cells of the apical meristem also produce parenchyma cells that make up the root cap. The root cap shields the apical meristem. It secretes a rather slimy substance that acts like a lubricant, enabling the root tip to push through the soil more easily. Because the root cap is at the forefront of all downward growth, it goes through a lot of wear and tear. Root cap cells constantly are damaged or sloughed off—and just as constantly, the apical meristem replaces them.

In the region of elongation, cells start elongating and forming into specialized tissues. This is the area where sieve tubes start forming for phloem and where vessel elements for xylem start to form. By the time cells are pushed upward into the region of maturation, epidermal tissue has formed, and root hairs are about to form. In plants, some of the epidermal cells send out extensions called root hairs. These root hairs spread out over a large surface area and increase the plant's capacity to take in water and nutrients.

Ground tissue has also formed by the time cells reach the region of maturation, and the vascular cylinders are about to form. A vascular cylinder encloses the conduction cells and consists of primary xylem and phloem. The cylinders also contain one or more layers of parenchyma cells, called the pericle.

Vascular cylinders divide the ground tissue system into cortex and endodermis regions. The cortex is located just inside the epidermis. This is the largest region of the developing root, and for a good reason: It's responsible for much of the transport within the plant. The cortex is made up of loosely packed parenchyma cells. The innermost section of the vascular cylinder is the endodermis, also known as the pith region. The endodermis is a layer of specialized cells that regulates what enters the center of the root. Each cell in the endodermis is encircled by a narrow band of suberin (the same stuff that's in cork cells), which in this instance is referred to as the Casparian strip.

Substances that are absorbed by the root hairs enter the root and move around freely toward the root center. When these substances reach the endodermis, the Casparian strip blocks their passage, forcing them to pass through the semipermeable cell membranes of the endodermal cells. The cell membrane regulates the transport of substances into the interior vascular tissues of the plant.

Stems

Stems come in all shapes and sizes. Strawberry plants produce stems that crawl along the surface of the ground. Cacti produce thick, fleshy stems—in fact, what you think of as a "cactus" is actually just a giant stem. Crocuses have underground stems called corms that specialize in storing food. Even potatoes are actually plant stems that have been modified to store food for the rest of the plant.

Like roots, stems have their own meristems. During growing season, only the portions that have meristems will grow. Stems grow only at their tips because that's where the apical meristem is. Stems also grow in circumference, due to lateral meristems. No other parts of the stem are growing. When you nail a sign on a tree, it will stay there. The sign may move outward, thanks to lateral meristems, but it won't move upward.

Stems grow much the same way that roots do, only above ground. And, of course, the apical meristems of shoots don't need a root cap (or "shoot" cap); they're not pushing their way through soil. In shoots, the apical meristem pushes upward, while beneath it, the cells divide at different rates and in different directions, rapidly differentiating in shape, size, and function. The same three tissue systems that form in roots—ground, dermal, and vascular—also form in shoots. Likewise, the cortex and endodermis, or pith, that developed in the vascular tissues of the roots will continue into the shoots.

As the shoot lengthens, bulges of tissue start to appear on the sides of the apical meristem. Each of these bulges—called *nodes*—is a rudimentary leaf. As growth

Weird Science

Unless tree roots manage to buckle a sidewalk or close off a sewer drain, most of us don't pay much attention to them. But the ground you walk on *is* riddled with roots, most of them meters beneath you. Even simple root systems manage to penetrate the soil to a depth of 2 to 5 meters. In hot deserts, root systems go much deeper. One mesquite bush is known to have sent its roots down 53.4 meters (175 feet). Even when roots don't grow deep, they grow extensively. One 4-month-old rye plant grown in 6 liters of soil water produced roots that, if laid in a sheet, would have covered 600 square meters.

Bio Buzz

The place on a plant stem where a leaf attaches is called the **node.** The stem region between two nodes is called the **internode.**

continues, the stem progressively lengthens between tier upon tier of new leaves. The stem region between two nodes is called an *internode.*

Nodes can develop on any part of a stem, as long as the apical meristem is present in that portion. Leaves develop only in that portion of the plant that is currently growing. When the primary body of a plant is forming, the dermal, ground, and vascular tissues of the stem organize themselves in distinctive ways. Usually, the primary phloem and xylem develop within the same sheath of specialized cells, called vascular bundles. Vascular bundles are multistranded cords that develop lengthwise through the ground tissue system of a plant's primary and lateral shoots.

In dicots, vascular bundles are arranged in a ring, dividing the ground tissue into cortex and pith. In most monocots and some dicots, the vascular bundles are scattered throughout the ground tissue. In dicot stems, the cortex is the region between the epidermis and the vascular bundles. The pith is the stem's center, inside the ring of vascular bundles. Xylem is located near the inside of the stem, while phloem is located closer to the outside. In monocots, both xylem and phloem are located near the center of the stem.

Try It Yourself

Over the centuries, science has advanced because people thought and pondered things of interest. Many questions are not immediately answerable, and some may never be. Put on your thinking cap and see if you can advance the scientific cause.

Give some thought to all the different environments on Earth: humid jungles, grassy prairies, hot deserts, frosty tundra, and so on. Pick an environment, think about what you know of plant structure, and design a flowering plant that would do well in that environment.

Cacti live in one of the harshest environments on earth: very dry, very hot climates with poor, sandy soil. Yet, they are a vascular plant. What type of evolutionary adaptations might cacti have gone through to be able to survive in deserts?

Stems have three primary functions: support, transport, and storage. Stems are the primary means of upward support for any plant, allowing it to expose its photosynthetic cells to sunlight. The vascular bundles enclosed within stems are also the primary means of transporting water and nutrients, obtained through the root system, to all the above-ground portions of the plant.

In some plant species, the stems are specially adapted for storage. As we've already noted, potatoes are a great example of stems as "pantries." The stems of many plants are full of parenchyma cells, which are excellent for storing food and water for later use. Cactus, in particular, have very fleshy stems that retain water for later use.

Leaves

Leaves are photosynthetic specialists: Each leaf is a photosynthesis factory. Roots and stems gather and transport water and nutrients, but leaves produce energy. (See Chapter 4, "Understanding the Cells: The Building Blocks of Life," for a review of this fascinating process.) Most leaves of vascular flowering plants are thin and flat, but like other plant parts, leaves come in all shapes and sizes.

Take a cactus, for example. Because we've already established that what people consider a cactus is really a stem, what do you think the leaves on a cactus look like? They're actually the prickly spines. In the case of cacti, little photosynthetic activity takes place in the spines; most of it occurs in the stems.

The tendrils found on pea plants and ivy are also a type of leaf. Again, not much photosynthesis occurs in tendrils because plants with tendrils also have "real" leaves. Tendrils tend to wrap themselves around objects, pulling the plant ever upward toward sunlight.

Then there are the leaves of carnivorous plants such as the pitcher plant, a plant whose leaves are shaped just like a pitcher. (And like a true pitcher, the leaves of pitcher plants hold water, eminently useful for drowning any insects that fall in.) Nearly 500 species of carnivorous plants exist in the world. Charles Darwin referred to the leaf systems of some of these plants as "temporary stomachs" because they act almost like a human stomach. The leaf systems of carnivorous plants actually secrete digestive juices. Most carnivorous plants live in soil that is relatively low in nitrogen and phosphorus—places such as wet, acidic bogs.

Leaves can be round, needlelike, heart-shaped, or rectangular—almost any shape you can imagine. Their edges can be smooth or serrated. Leaves can be virtually microscopic, or meters in width. Despite the diversity in shape and size, all leaves are composed of ground, vascular, and dermal tissues.

Bio Buzz

Mesophyll is the ground tissue of plant leaves. It is composed of chloroplast-rich parenchyma cells.

The broad, flat portion of a "typical" leaf is called the blade and is where most photosynthesis takes place. Leaf blades are attached to the plant's stem by a stalklike structure called a petiole.

In most leaves, the epidermis, or dermal tissue system, is a single layer of cells coated with a waxy, impermeable cuticle. Everything that enters and exits the leaf—oxygen, water, carbon dioxide—does so through stomata in the epidermis.

Photosynthesis occurs in the ground tissue system of leaves, a region referred to as the *mesophyll.*

The vascular tissue system in leaves takes the form of vascular bundles, such as those found in other parts of the plant. In the case of leaves, the vascular bundles are referred to as veins. Leaf veins are embedded in the mesophyll. The primary function of leaves is to take in sunlight, carbon dioxide, and water, and convert that into sugars, which the plant then uses as an energy source.

For all the diversity, angiosperms (what we typically recognize as "plants") are remarkably similar in their composition. They're all organized differently, but it still boils down to shoots and roots.

Although we've covered only the bare basics of plant structure in this chapter, it gives you an idea of just how diverse plants can be. We've talked about the structural tissues of plants, how they grow, and the different components of plants—namely, roots, stems, and leaves. In the next chapter, we'll be talking about the world of woody plants, commonly referred to as trees.

The Least You Need to Know

➤ Seed-bearing vascular plants can be divided into two main categories: gymnosperms (which bear cones), and angiosperms (which bear flowers).

➤ Angiosperms are the most numerous plants on Earth. Angiosperms also can be divided into two main categories: monocots (such as corn and grasses), and dicots (such as roses, maples, and lettuce).

➤ Plant tissues are organized into three main organ systems: roots, stems, and leaves. All three tissue systems can be found in each organ system.

➤ Plants do not grow all the time; during a growing season, plants grow only through meristems. Apical meristems cause plants to lengthen, and lateral meristems cause plants to increase in circumference.

Chapter 19

Woody Plants

In This Chapter

➤ Herbaceous and woody plants

➤ Tissue formation of secondary growth

➤ Bark and wood

➤ How far can you grow?

➤ The autumn leaves

It's easy enough to think of the plants in your garden as "plants," but the plant kingdom contains a lot of different species, nearly all of which you know as plants, but many of which you think of as something else.

Take trees, for instance. Obviously, they're trees, but they're also flowering plants. Maples, elms, oaks—they're all woody flowering plants.

In this chapter, we'll talk about what makes a tree actually a tree. Why do some flowering plants become towering trees, while others grace the flower pots on your patio? We'll discuss how wood forms, bark develops, and leaves flame into the colors of autumn.

Herbaceous vs. Woody Plants

Flowering plants have life cycles that start with germination, go on to seed formation and development, and end in death. Because of the enormous diversity of the plant kingdom, there are lots of ways to classify plants. One way is by their life cycle.

Another way is by their basic body makeup. *Herbaceous* plants have stems that die on a yearly basis, while the stems of *woody plants* develop the hard, supporting tissue we recognize as wood. On many points, these two methods of classification overlap.

Annual plants are some of the most common herbaceous plants. They complete their entire life cycle in a single growing season. Consequently, the name itself, annual, is self-explanatory. In temperate climates, this means that the plant will germinate, form seeds, develop its entire body structure, and die all within the space of about six months. Geraniums are annual plants.

Biennial plants are also a common type of herbaceous plant. Biennial plants live for two growing seasons. During the first growing season, biennial plants develop roots, stems, and leaves. During the second season, the flowers and seeds develop. At the end of the second growing season, after flowers and seeds have formed, the plant dies. Carrots and snapdragons are biennial plants.

Perennials are plants that continue to grow season after season. Theoretically, under correct conditions, they'll continue growing indefinitely. In perennial plants, seed development and vegetative growth occur (or reoccur) year after year. Perennials can be herbaceous plants or woody plants.

Herbaceous perennial plants reappear each year from their roots. The shoots and leaves of perennials die back, but the roots, while not really active, remain alive. When conditions are right, the apical meristem will produce shoot growth. Herbaceous perennials can grow bigger every year, although there are limits. They usually don't get much bigger in length or height (when was the last time you saw a delphinium that was 5 meters tall?), but they can increase slightly in circumference.

Woody plants are perennial plants, but most of us just think of them as trees. Not all woody plants are trees, but most of them are. All gymnosperms (such as conifers) are woody plants, as are many monocots (such as palm trees) and dicots (such as maples and elms).

Bio Buzz

A **herbaceous plant** is any seed plant whose stem withers and dies on a yearly basis. A **woody plant** is any seed plant whose stem develops a hard, durable tissue that we call wood.

Try It Yourself

It's easy to confuse biennials with perennials, particularly if all you're considering is a blooming plant in a garden. Because biennials reseed themselves, they frequently come up year after year in the same spot. Although the plant is always there, in this instance, the distinction between biennials and perennials is that the perennial that reappears every year is always the same plant. A biennial that appears every year is a completely different plant.

Weird Science

With a foot-thick stem and heights of 120 feet, bamboo can look like a tree, but it is actually a grass. It's extremely long-lived; bamboo forests in China can live for more than 100 years. Yet when it's "time" for a bamboo forest to die, the whole forest just keels over. Bamboo reproduces asexually from underground rhizomes, but it also reproduces sexually at regular intervals of 30, 60, or 120 years, after which it dies. These die-offs of entire bamboo forests frustrate farmers, but they are devastating to endangered pandas, who depend on bamboo for food. Recently, scientists have discovered a way to make bamboo flower in the laboratory, a technique that holds promise for producing seeds to replenish bamboo forests at regular intervals.

Herbaceous and woody plants possess the same tissue systems and structure. Both have roots and shoots. Both have ground, vascular, and dermal tissues. And both have xylem and phloem. What really sets herbaceous and woody plants apart is the formation of secondary tissues, or secondary growth.

Tissue Formation of Secondary Growth

When woody plants are young, their stems and roots are very similar to those of herbaceous plants. A young oak seedling, just emerged from the soil and reaching upwards for sunlight, doesn't look much different than a daisy seedling. They're both small, green, and flexible, and they both have leaves. The differences between the two won't emerge until after the oak's lateral meristems become active and start producing large amounts of secondary tissue, particularly secondary xylem. The massive woody stems that we recognize as trees (and their equally massive roots) originate in a lateral meristem called the *vascular cambium*. It's this tissue that really sets woody plants apart from herbaceous plants.

Bio Buzz

The **vascular cambium** is a lateral meristem that increases the circumference of plant stems and roots.

In some perennial plants, the vascular cambium has sprung to life every growing season for hundreds or even thousands of years. This kind of lateral meristematic activity produces the gigantic redwoods found in northern California. Some of these trees tower more than 110 meters above the ground and have a circumference of more than 30 meters, all thanks to the secondary growth produced by their vascular cambiums.

In stems, primary growth resumes every spring at the buds, or places where new twigs and leaves will emerge. At the same time, secondary growth proceeds inside the stem, at the vascular cambium. When the vascular cambium is fully formed, it is a complete cylinder encircling the stem that is the thickness of only one cell or a few cells. Out of these few cells, the circumference of a tree can increase for centuries.

Some cells in the vascular cambium will give rise to secondary xylem and phloem. The secondary xylem and phloem arise on either side of the vascular cambium, next to the primary xylem and phloem, respectively, and extend longitudinally throughout the plant, just like the primary xylem and phloem. Without the formation of secondary xylem, secondary phloem, and rays, water and nutrients wouldn't be able to travel up, down, or sideways in the ever-increasing woody stems.

Other cells in the vascular cambium produce rays of parenchyma cells. These rays are vascular tissue that deliver water sideways through the stem. The rays produce a pattern that looks like sliced pie. You can see this pronounced radial pattern in the stump of any tree that has been cut down.

The vascular cambium rings the tree. Secondary xylem forms on the inner side of the ring of vascular cambium, and secondary phloem forms on the outer face of the ring. As the secondary tissues accumulate, often over hundreds of years, it seems like the secondary growth would eventually squash the vascular cambium right out of existence.

Try It Yourself

If you want to see the largest tree trunk in the world, head for Sicily on your next vacation. The tree that holds the record for lateral meristem activity in its vascular cambium is a chestnut tree growing in Sicily. The circumference of the base of the tree is more than 58 meters.

Not so! As the inner core of secondary xylem gets thicker and thicker, it steadily displaces the cells of the vascular cambium and pushes them forward, toward the surface of the stem. In turn, the vascular cambium remains in place (encircling the entire tree) by dividing sideways. In this manner, the cells of the lateral meristem remain in place, in an ever-increasing circle.

In woody plants, the development of secondary tissues by the vascular cambium is obvious: The tree keeps getting bigger and bigger every year. But at the same time that secondary xylem and phloem are developing in the stem, they're also developing in the root system. The roots of woody plants also increase in circumference every year, thanks to the vascular cambium lateral meristem.

All this lateral meristem activity takes energy—are there any real advantages to having woody stems and roots? The advantages for trees are the same as for any plant:

By growing higher, they get more sunlight. In the case of trees, they've just developed this principle to the max. Even the smallest, youngest tree usually towers over the largest stemmed nonwoody plant. Trees have got the metabolic means to grow large shoot and root systems, so they're better able to compete in almost any habitat.

We've been discussing the vascular cambium of woody plants, but nonwoody plants also have a vascular cambium that increases the circumference of their roots and stems, particularly in perennial plants. It just operates on a much smaller and less obvious scale.

Try It Yourself

If you ever find yourself in the woods or in the desert without water, you won't die of thirst if there are plants around. Plants tend to be full of water, especially plants with a lot of xylem. Vanilla plants and philodendrons in particular rank as good choices. If you're lost in the desert, you can always rely upon cacti to provide water, although the water found in most cacti is slimy, bitter, and, of course, hot.

Arf! Arf!: Bark and Wood

No matter how old the tree, secondary phloem ends up restricted to a relatively small zone just outside the vascular cambium. The secondary phloem consists of sieve tubes and thin-walled living parenchyma cells, interspersed among bands of thick-walled fibers. Only the sieve tubes close to the vascular cambium remain functional; the rest of the phloem is dead and helps to protect the living cells.

Bio Buzz

Bark is composed of all the tissues that are external, or located on the outer side, of the vascular cambium. **Periderm** and secondary phloem are the major tissues found in bark.

It's the secondary xylem that accumulates like mad. In trees in which secondary growth is extensive, xylem typically makes up 90 percent of the tree. As growing seasons pass, the inner core of xylem continues to expand outward. As you might guess, this expansion produces a great deal of pressure, most of it directed toward the stem, or root, surface. Eventually, this pressure ruptures the cortex and outer surface of the phloem.

Courtesy of this rupturing, parts of the cortex and epidermal tissues split away, and a new surface must form. This new surface, the *periderm*, forms out of the cork cambium. The periderm and secondary phloem are the major components of *bark*.

The periderm itself consists of several tissues: cork, secondary cortex, and the actual cork cambium that produces these tissues. Shortly after the formation of the vascular cambium, the outermost parenchyma cells (either roots or shoots) form the cork cambium. Because the cork cambium contains parenchyma cells, it (and, by inference, the periderm) retains the capacity to divide.

Cells dividing in the cork cambium produce cork toward the outside and living ground tissues toward the inside. Only the cork cells closest to the cork cambium remain alive because they're the only ones with access to water and nutrients from the xylem and the phloem. The rest of the cork cells die and can't expand. But the continuous production of secondary xylem and phloem keeps splitting the cork and moving outward.

Because of its fatty suberin layers, cork provides protection, insulation and waterproofing for the stem or root surface. Cork also forms over wounds, helping to protect the tree from invasion.

Since the cells in plants are living—even in woody stems and roots—they need to inspire and respire oxygen and carbon dioxide. How do gases exchange across a waterproof layer of cork? The gases exchange through *lenticils*.

You can see lenticils yourself by looking at the cork from a wine bottle. The dark spots and streaks sprinkled throughout the cork are all that are left of the lenticils.

Think About It!

If you've ever wondered why the bark of common trees looks so rough, it's because its cork cells continually rupture, creating an uneven outer surface on many trees.

Think About It!

Thanks to cork, most bark can heal wounds inflicted upon the tree. But sometimes the wound is too extensive. If you've got a tree that has lost a lot of bark, it's a good idea to seal the stripped area with tree sealant (available in most hardware and gardening stores). The sealant won't help generate any new cells, but it will keep out invaders, giving the tree time to recuperate on its own.

Changing Wood: Heartwood and Sapwood

As trees age and change in appearance, the function and appearance of the wood itself change. Heartwood accumulates in the center of stems and roots. Heartwood is a dry tissue, in that it no longer transports water or nutrients. Heartwood is also a dumping ground for metabolic wastes created by the tree. Gums, tannins, resins, and oils are just some of the metabolic wastes deposited in heartwood. Eventually, all this metabolic waste clogs up the oldest xylem tubes. This clogging by metabolites darkens heartwood and makes it aromatic. Heartwood also strengthens a tree, offering structural support and enabling the tree to grow to enormous heights.

Sapwood is almost the exact opposite of heartwood. Sapwood is wet, pale, and weak. Within the tree, sapwood is located between the heartwood and the vascular cambium. Sapwood conducts plenty of water and nutrients. Maple trees possess some of the best known sapwood. Every spring, across the northern woods of North America, tree farmers insert metal tubes into the sapwood of sugar maples. The resultant sugar-rich fluid produced by the secondary phloem, called sap, gets turned into maple syrup.

Early? Late? Hard? Soft?

In the temperate regions of the globe, the vascular cambium isn't active all year round, but only during the growing season. The vascular cambium is inactive during cold winters and dry spells. When the first xylem cells are produced at the start of each growing season, they tend to be large with thin cell walls. These are the cells that form *early wood*. As the summer days heat up and dry out, the vascular cambium produces smaller cells with thicker cell walls. These are the cells that form *late wood*.

When you look at a transverse section of a tree trunk, the difference between the cells sizes of the early and late wood creates alternating bands that reflect light differently. These differences are called growth rings. In areas where seasonal changes are predictable, growth rings can be used to determine the age of the tree. Trees growing in temperate regions usually produce only one ring a year. But in areas where the seasonal changes are unpredictable, the growth rings are a reflection of just that: growth. The growth may or may not have occurred within the chronology of a single year.

In deserts, trees grow only when there is water. So, even if the region is temperate, if it's also dry, the tree rings will reflect the presence of water. If a thunderstorm roars through a dry area, the trees may reflect that additional water with an additional

Bio Buzz

Lenticils are localized areas in the cork layer where the cells are loosened up a bit, somewhat like stomata. Gases are able to pass through these open areas of the cork layer.

Weird Science

Because heartwood doesn't conduct water or nutrients, trees can get along without it. Before California's redwoods were protected, some loggers in the early 1900s obviously knew this. They're the ones that cut tunnels through some of the giant redwoods big enough to drive a car through. Because the majority of the tunnel excavation was composed of heartwood, the trees didn't die.

ring. Because seasonal change is almost nonexistent in the tropics, growth rings usually aren't even a feature of tropical trees.

Think About It!

Different types of trees absorb and accumulate nutrients differently. Consequently, the accumulated resins and oils will be different in different tree species. You can experience these distinctions yourself with your own backyard barbecue. The next time you're cooking, get some chips of different wood: mesquite, cherry, maple—almost any type will do. Toss different chips on the coals, and observe the difference in smoke and smell, as well as the taste of your food.

Bio Buzz

Early wood is composed of the first xylem cells produced during the growing season. Early wood cells are large and thin-walled. **Late wood** is composed of xylem cells formed further into the growing season. Late wood cells are smaller and thick-walled.

Xylem cells also determine whether a tree is classed as a hardwood or a softwood. Although it's not a hard-and-fast rule, temperate dicot trees tend to be hardwoods, while gymnosperms tend to be softwoods. The size and composition of the xylem determines whether a tree is "hard" or "soft." Hardwood trees, such as oak, hickory, and maple, have xylem that's composed of fibers, tracheids, and vessels. The vessels may occur only in early wood, or they may occur in both early and late wood—it depends upon the species of tree. Softwood trees, such as pines and conifers, have tracheids and parenchyma rays in their xylem, but no fibers or vessels.

Hardwoods are used in the housing industry and for any building project that requires strength and durability. Not only are hardwoods strong, but they rarely warp, particularly if they are cut and dried correctly. Thus, hardwoods are premium lumber, and because trees can be replanted and renewed, they're an excellent resource. Unfortunately, the best hardwoods are not fast-growing; they take a long time to achieve maturity.

Softwoods grow faster than hardwoods, so they're ideal candidates for recycling and regrowth. Because of their softness, softwoods aren't useful for housing or building projects, but they're often used in furniture, especially pine. Without fibers in their xylem, pines, spruces, redwoods, and other conifers are genuinely weaker and less dense than hardwood trees. Their wood is "soft," hence the name softwoods.

How Far Can You Grow?

If the vascular cambium reactivates at the start of every growing season, can trees just keep growing forever? Theoretically, yes, but usually, no. Under perfect conditions, many trees could just keep growing—look at the bristlecone pines and the California redwoods. But unlike humans, trees cannot run away from approaching disaster, nor do they have an immune system to help fight off bacterial, viral, or protist invasions.

Some trees manage long lives because they live in habitats that are too harsh or too remote for most invaders, and their lateral meristems just keep activating year after year. But the majority of trees live in habitats where they are sitting ducks, so to speak. Humans, bugs, and bacteria are all present. These trees can still manage to live long lives through *compartmentalization*.

Bio Buzz

Compartmentalization is the process by which a tree deploys compounds and walls off invaders.

Even without an immune system, most trees can limit their vulnerability by constructing a fortress of thick cell walls around wounds (and stationary invaders), and deploying toxic compounds such as phenols. All these responses taken as a whole are called compartmentalization.

Compartmentalization works well if the attack is not too extensive. The tree thickens its cells and closes off the invader from the rest of the tree. Then it produces new tissues over the wound or infection site, effectively controlling the damage. But if the attack is extensive or the tree responds too late, the tree will die, not only from the invasion, but also from its own response.

Compartments do more than just wall in an invader; they also close off transport systems and living tissue. If the tree is forced to make too many compartments, the compartments will block the flow of nutrients and water to other parts of the tree, causing it to die. Some toxins produced by trees in response to an invader are so toxic that they not only kill the invader, but they also kill the tree itself. So, although trees theoretically could grow forever, outside influences usually bring them down.

From roots and shoots, from sap to bark, trees go through some amazing changes every year—more so than many animals. Through all their growth and change, trees are still one of the best renewable resources on the planet.

Trees follow many of the same growth patterns that any vascular plant follows (shoots, roots, stems, leaves, and so on), but it's the formation of secondary xylem—otherwise known as wood—that really sets these plants apart. In the next chapter, we'll be talking about the tissue systems that plants have developed to obtain nutrients and water.

The Least You Need to Know

➤ Flowering plants frequently are classed as either herbaceous or woody. Herbaceous plants whither and die on an annual basis. Woody plants are characterized by the development of secondary tissues that we recognize as wood.

➤ Herbaceous plants can be either annual (the plant completes its entire life cycle in one growing season), biennial (the plant completes its life cycle in two growing seasons), or perennial (the plant has an indefinite life cycle, reactivating during the growing season).

➤ Woody plants are characterized by the presence of an active vascular cambium, a lateral meristem that produces secondary tissues such as secondary xylem and secondary phloem. These expanding secondary tissues are responsible for the ever-increasing girth of most trees.

➤ Bark forms when the outward pushing and subsequent stress of the secondary tissues and vascular cambium rupture the cortex and the outer surface of the phloem.

➤ Trees have both heartwood and sapwood. Heartwood, which is hard and dry, is found in the center of the tree and is used for storage of metabolic wastes. Sapwood is soft and full of water, and is located between the heartwood and the vascular cambium.

Chapter 20

Plant Nutrition and Transport

> **In This Chapter**
>
> ➤ Nutrition requirements
>
> ➤ Where are the minerals?
>
> ➤ Water and mineral absorption
>
> ➤ Water transport and conservation
>
> ➤ The phloem highway

Plants are masters of adaptation. Take a look around; nearly every spot on the globe is capable of supporting plant life. No matter what the climate or soil conditions, plants find a way to survive. Nevertheless, all plants have basic water and nutrition requirements that have to be met for survival. If you see a plant growing somewhere—no matter how dry, cold, or desolate the location—you can rest assured that the plant is meeting its nutritional requirements.

In this chapter, we're going to talk about how all plants develop the specialized tissues that enable them to meet their nutritional and water requirements. We'll discuss what nutrients plants need to grow and how they deliver the goods to every cell.

Nutrition Requirements

Nearly all plants are phototrophs in that they use the energy from sunlight to synthesize the organic compounds necessary for life, such as water, carbon dioxide, and some minerals. But just like people, plants don't have an unlimited supply of the resources they need to stay alive. In fact, for all their proliferation, plants have a rather tough time of it. Out of every 1 million molecules of air, only about 350 of those molecules are carbon dioxide, which is a major plant requirement.

Plants also spend their life rooted to the spot, literally. Soil conditions can vary drastically and frequently, leaving the plant to either deal with the new environment or die. Plants that live in marshes and swampy bogs must cope with too much water (the soil in a swamp or bog is never dry) and too few nutrients (the soil in a swamp tends to be very acidic, with virtually no nutrients that plants can use). Hence the evolution of carnivorous plants, which have adapted and evolved specialized structures that enable them to get the nutrients they need, regardless of the nutrients in the soil. But too much water is not the norm for most plants. Most plants exist in soils that are frequently too dry.

Try It Yourself

You can see plant xylem in action right in your own kitchen. Trim off the bottom of a large piece of celery, and place the celery in a glass of colored water. (Red is particularly good.) Now all you have to do is wait. Within a few hours, the celery stalk will start pulling the water out of the glass and into its xylem tubes. Because the xylem tubes on celery are especially prominent, you'll see the colored water moving up to the top of the celery stalk.

Bio Buzz

A **nutrient** is any element that is essential to an organism because, either directly or indirectly, that element plays a role in the organism's growth and survival that no other element can fulfill.

Only in well-tended and well-fertilized gardens do you see picture-perfect plants. They're the plants that get consistent and appropriate doses of water and sufficient amounts of dissolved minerals. The rest of the plant kingdom makes do with what it has. Fortunately, this making do is primarily responsible for the diversity of the plant kingdom. Most aspects of plant structure and function are a response to low concentrations of essential nutrients or to other environmental factors.

All living organisms need nutrients to survive, but what exactly do we mean by the term *nutrient*? When it comes to plants, all of them need oxygen, hydrogen, and carbon. Plants obtain these three elements through the photosynthesis of water and carbon dioxide. These elements are the basics of plant nutritional requirements.

But there are at least 13 other elements that plants need to thrive and survive. Generally, these elements are dissolved in soil water in forms that are capable of binding to clay. The nutrients that plants need for survival are divided into two general categories: macronutrients and micronutrients.

Minerals, Minerals—Where Are the Minerals?

Plants obtain most of their required minerals from the soil. Soil consists of particles of minerals mixed with varying amounts of decomposing organic material. The minerals come from the weathering of the Earth's hard rocks, and the organic material comes from decaying organisms and the litter they create during their lifetimes, such as feces and leaves.

Soil cloaks the surface of most land habitats. The proportion of minerals found in any given section of soil varies from region to region, and even within sections of the same habitat. Mineral particles vary widely, as does the extent to which they are compacted in the soil. Typically, soil is classed into three very rough, general categories, based upon the type of particles found in the soil:

1. Sand, such as you find at the beach. As far as soil goes, sand particles are fairly large; each grain is about 0.05 to 2 millimeters across.

2. Silt, such as you find at the bottom of ponds. Silt particles are much smaller than sand, with individual particles being about 0.002 to 0.05 millimeters across.

3. Clay, which has the finest particles of all, smaller than sand or silt.

The composition of soil greatly influences the type and volume of plant life that can be supported in any given habitat. If the soil becomes gummy when wet, the spaces between the particles compact too tightly, leaving no air spaces. If the soil dries quickly when wet, the particles aren't compacted enough, so they're unable to retain water. The variations of soil particles are endless—another reason why only pampered garden plants tend to be "perfect."

Think About It!

If you want your garden to thrive, you might try adding some humus to the soil. Decomposing organic material, known as humus, is another critical ingredient of soil. Humus absorbs water, releasing it slowly. Humus also has an abundance of negatively charged organic acids, which bind to dissolved mineral ions. Where humus is concerned, though, you *can* have too much of a good thing. Soil with 10 to 20 percent humus is optimum.

Bio Buzz

Loams are soils with roughly equal proportions of sand, silt, and clay.

Sand, silt, and clay all have critical advantages for plants. But when it comes to plants and soil, the right mixture of all three, called *loam*, represents the healthiest environment for plant growth.

If the mix of soil particles isn't just right, plants will still grow in it, but they won't necessarily thrive. Additionally, a soil mix that isn't right can create other problems for both plants and humans. In soil with too high of a sand content, leaching occurs. Leaching is the removal of nutrients by water percolating through the soil. Sand can't bind minerals the way clay can, so the minerals leach out of the soil. In soil where leaching has occurred, crops don't grow well (if at all), and any crops that do grow lack nutrients, which humans need through plant consumption for general health.

Even when the soil mix is correct, the geographical terrain can cause problems for plants. Erosion (the movement of land by the forces of wind, water, and ice) removes tons of topsoil and the minerals contained therein every year. Erosion is particularly pronounced in areas where gravity gives water a helping hand, such as on hills.

Think About It!

If you'd like to find out the composition of the soil in your area, many local colleges offer this service. You can also check out the U.S. Department of Agriculture's Web site, at http://www.usda.gov/usda.htm, for other interesting information and services.

As you can see, minerals dissolved in water are critical to a plant's health. But once the minerals are in the soil and are available, how does a plant get them? That's what we'll talk about next.

Water and Mineral Absorption

The root systems of plants act like miners. They venture out, usually deep into the ground, probing and searching for minerals. In terms of energy expended, mining ventures by plants are an expensive proposition. Plant bodies are stuck in one place, forced to expend enormous amounts of energy to develop a root system capable of garnering the resources they need.

Furthermore, as a plant's root system extends through the soil, it encounters different types of soil particles, as well as some formidable obstacles, such as rocks and other root systems. Every time a plant encounters a new texture or different composition in soil, it must form new roots and branch off in a new direction.

It's not so much that plants are exploring the soil looking for minerals. It's more a matter of trying to simply encounter soil with the correct mix of minerals and ions. When this encounter occurs, the concentration of nutrients stimulates the root system to greater growth. Encountering minerals gives plants the veritable boost. It becomes a never-ending circle.

Think back to the root systems of plants, all those vascular tissues, endodermis, pith, root hairs, and Casparian strips. Recall also the movement of fluid and permeable membranes. All components of a plant's root system are used to seek, garner, and transport the water and minerals found in soil.

The water molecules found in soil are only weakly bound to the clay particles. It's very easy for the water molecules to break away and move across the cells of the root epidermis and into the root's *vascular cylinder.*

The vascular column is surrounded by a layer of cells called the endodermis. The Casparian strip runs through the abutting cell walls. If you recall, the Casparian strip is waxy, and water molecules are unable to penetrate it. For water to get into the vascular cylinder (and on to other parts of the plant), it must move into, pass through, and then exit the endodermal cells.

After water molecules have permeated the walls of the endodermal cells, they encounter transport proteins with the cell's plasma. These proteins allow some solutes to pass through, but not others. Basically, the transport proteins act like border cops that screen out undesirable substances from the water molecules.

The function of the transport proteins in the endodermal plasma goes beyond simple screening. The proteins control and regulate the types and amounts of solutes that the root system absorbs from the soil water.

Plants would not be able to meet their water requirements without root hairs. Root hairs are extensions of specialized epidermal cells. Their presence increases the surface area of the soil that's available to absorption. Whenever a plant is engaged in primary growth—which is during much of the growing season—its root system usually is developing millions or billions of root hairs.

Bio Buzz

The **vascular cylinder** is the central column of vascular tissue found in the root.

Weird Science

Vascular plants go through huge amounts of water every day. A single adult corn plant sucks in more than 3 liters of water in a single day. Take a guess how much water a full-grown oak processes on any given summer day. Hard to know, but probably hundreds of liters!

Bacteria and fungi also play a key role in a plant's water absorption. Usually, it's a symbiotic relationship between the plant and the bacteria. Think back to the carnivorous bog plants. They're carnivorous because they need nitrogen, which is not available in

their soil. Nitrogen is critical to plants, so much so that if their roots can't find any, they'll eat insects to make up for the lack. Nitrogen is abundant in the air, but plants aren't capable of nitrogen-fixing; certain types of bacteria, however, are. If you recall from Chapter 6, "Watch Out! Viruses and Bacteria," nitrogen-fixing is the process by which bacteria convert gaseous nitrogen into useable ammonia compounds.

The root systems of many plants have developed symbiotic relationships with nitrogen-fixing bacteria. The bacteria convert gaseous nitrogen into forms that they—and the plant—can use. The bacteria use up some of the organic compounds found in the plant's tissue, but they provide plenty of nitrogen in return.

Plants also form symbiotic relationships with fungus. Think back to Chapter 8, "Fungi!", and recall mycorrhizae, the symbiotic relationship between fungi and young roots. Fungal filaments (hyphae) penetrate root cells or form velvety coverings over the roots. The hyphae create an even greater surface area, allowing the plant to absorb more minerals, as well as scarce minerals that it couldn't absorb on its own. In turn, the hyphae obtain sugars and nitrogen from the plant's roots.

Water Transport and Conservation

Root hairs, bacteria, and fungi all help plants obtain water and minerals. Once the water and solutes are inside the plant, the endodermal cells in the vascular column screen and regulate the solutes. So where does the water go from there?

Scientists used to think that the water that was absorbed by a plant's root system simply pushed water up into the plant. In very small plants, such a simple system could work. But what about giant redwoods that are more than 100 meters high? How does the water get from the ground (and some of those roots are tens of meters deep into the soil) and up to the topmost branches? In a word: evaporation. Evaporation is the primary force that moves water from the tiniest root hairs up to the tip-top leaves.

The interior of a plant is fluid in motion. Yet, for all the water that moves through a plant, that's exactly what it's doing—moving through the plant on its way out. Only about 2 percent of all the water taken in by a plant is actually retained by the plant and used for photosynthesis and metabolic processes. More than 90 percent is lost, usually through stomata.

The daily life of a plant consists of water that's absorbed through its root system, water moving through the stem, and water flowing into the leaves. How does this delivery system work?

Transpiration

Within a plant's vascular tissues, water moves through xylem pipelines. That's what xylem does: transport water. If you remember from Chapter 18, "The Plant Body," xylem is composed of water-conducting cells called tracheids and vessel members. By maturity, both types of cells are dead, leaving only their lignin-reinforced (for our purposes, you need to know only that lignin is strong) walls behind. Because the conducting cells of xylem are dead, they can't be actively pulling the water through and upward.

Bio Buzz

The **cohesion-tension theory** states that the water in plant xylem is pulled upward to the leaves by the drying power of evaporation. This creates a continuous negative pressure within the plant that extends from the leaves all the way down to the roots.

For years, scientists puzzled over the transport of water within plants. Then a botanist, Henry Dixon, came up with a theory that accurately explained water transport in plants, the *cohesion-tension theory.* According to Dixon, the evaporation of water through the leaves and other parts of plants is a process called transpiration. There are three main points to Dixon's cohesion-tension theory of transpiration.

➤ The drying power of air (evaporation) causes transpiration. For water that is confined within narrow tubes of xylem, transpiration causes tension. This tension extends throughout the plant, from the leaves (where most evaporation takes place) all the way down to the deepest roots.

➤ The water within the xylem forms a fluid, unbroken column, which resists rupturing as the tension of transpiration pulls it upward. The water remains in a fluid column due to the strength of the hydrogen bonds between water molecules. The bonds are strong enough to stay together as the water is pulled upward.

➤ As long as water continues to escape from the plant's leaves, the water in the xylem remains under tension, pulling water from the roots upward to the leaves.

The hydrogen bonds between water molecules are strong enough to resist tension within the xylem. But the bonds are not strong enough to prevent the water molecules from rupturing during transpiration and escaping through the stomata.

Controlling Water Loss

Because more than 90 percent of all the water plants take in escapes into the atmosphere, controlling water loss is critical. Plants need only a tiny fraction of the water they take in to perform photosynthesis and other cellular functions, but that tiny amount of water is vital. If water loss exceeds water uptake for any extended period of time, the plant will wilt and die.

Plants depend on the soil for most of their water intake. But when it comes to controlling water loss, they're not entirely dependent upon the soil. Their *cuticle* and *stomata* are instrumental in keeping water loss at an acceptable level.

The surface of the cuticle is composed of deposits of water-insoluble lipids. The actual cuticle itself is made up of waxes embedded in cutin, a lipid polymer that is also impermeable. Underneath these waxes, cellulose threads crisscross the entire cuticle, giving it strength.

A cuticle severely restricts a plant's water loss, so much so that without a cuticle, most terrestrial plants would perish. Cuticles also restrict the inward diffusion of carbon dioxide and the outward diffusion of oxygen. But cuticles have no restricting effect on sunlight. Sunlight (needed for photosynthesis) passes easily through the cuticle.

However, because cuticles restrict the passage of gases, and because gases (such as oxygen and carbon dioxide) are critical to plants and the photosynthetic process, how do gases move in and out of plants? They do so through stomata, the cellular openings found primarily in the leaves of plants and also on stems. The opening of each stomata is defined by two highly specialized cells called guard cells. Guard cells control the opening and closing of each stomata. Stomata (and their guard cells) look a little bit like doughnuts: They appear circular and have a hole in the middle that opens and closes.

As pressure builds up within the plant, the force of this internal pressure causes the guard cells to bend to create a gap between them. Water escapes, the pressure drops, and the guard cells collapse against each other, closing the gap.

Because photosynthesis takes place during daylight hours, stomata remain open during the day in order to allow the passage of water and gases that occur

Bio Buzz

The **cuticle** is a waxy, water-proof layer of epidermal cells that covers plant regions that are exposed to air. The **stomata** are cellular openings in leaves that regulate the passage of water and gases.

during photosynthesis. This is when plants lose the most water, but they also gain the carbon dioxide they need for photosynthesis.

Stomata open or close depending upon the amount of water and carbon dioxide in the guard cells. As soon as the sun lightens the sky, photosynthesis starts. As the morning progresses, the carbon dioxide levels in all the plant's cells drop. This decrease in carbon dioxide triggers the transport of potassium ions into the guard cells.

But a decrease in carbon dioxide isn't the only thing that triggers the transport of potassium ions: So will blue wavelengths of light. As the sun arcs higher in the sky, blue wavelengths increase, triggering the movement of more potassium and more water into the guard cells.

By osmosis, water follows the potassium into the cells. This movement of water, triggered by potassium transport, creates the pressure that causes the stomata to open. When the sun goes down, carbon dioxide levels rise in the plant, and photosynthesis stops. First the potassium ions move out, then the water moves out, and, finally, the stomata close. Consequently, vascular plants conserve most of their water at night. During the day, the stomata are open, and water is evaporating into the atmosphere.

The Phloem Highway

Xylem conducts water and minerals, but critical as these are, plants do not live by water and minerals alone. They need nutrients, specifically the sugars and organic products of photosynthesis. Like xylem, phloem is a vascular tissue. But unlike xylem, the cells in phloem are alive at maturity. They can actively contribute toward the movement of nutrients; they don't have to rely on evaporation. Phloem consists of strands of parenchyma cells, fibers, and conducting tubes, all of which extend throughout the entire plant.

The conducting tubes in phloem are composed of *sieve tubes* and their corresponding *companion cells*. Companion cells are specialized parenchyma cells that help load organic compounds into the sieve tubes. Sieve tubes are the main components of phloem, the primary conducting tissue for nutrients. Companion cells and sieve tubes are positioned side by side, and both are positioned end to end throughout the plant to form the tubes that conduct sugars and organic compounds within the plant.

Storage and Transport of Organic Compounds

When leaf cells engage in photosynthesis, they create organic compounds. A small fraction of these compounds is used by the cells to fuel their own activity, but the majority of the carbohydrates created by photosynthesis is moved out to the plant's roots, stem, twigs, flowers, and fruit. Cells store most of these carbohydrates (the primary end product of photosynthesis) as starch in the plastids.

Before the plant can really use the storage products of photosynthesis, the cells must convert the storage forms of these organic compounds into smaller solutes that can travel through the phloem. Starch molecules are too large to transport across the cell membrane, and they're also too insoluble for transport to other plant parts. Likewise, proteins are also too large, and fats are too insoluble. The cells degrade the starch created by photosynthesis, eventually turning it into sucrose, an easily transportable sugar. Sucrose is the primary carbohydrate transported in phloem.

Translocation

Translocation is the technical name for the transport of sucrose and other compounds within plant phloem. High pressure drives the process of translocation. The pressure

involved in translocation is frequently five times that of the pressure in you car tires. Again, aphids provide a clue to the high pressure of the sap moving in phloem. When aphids tap into phloem, the pressure forces sap through the mouthpiece, through the aphid's gut, and out through the anal opening, where it forms a droplet of sugar water, or "honeydew."

Try It Yourself

If you park your car under a tree that's being invaded by aphids, you'll return to find it covered with sticky droplets, honeydew that has dropped off the aphids. What your car is really covered with is the sap of the tree's phloem; it just got there courtesy of the aphids, not the tree.

Pressure Flow Theory

Although phloem cells are alive, can cellular movement really account for the pressure found in phloem sap? Translocation works by moving organic compounds along gradients of decreasing pressure. The source of the initial pressure is not from the cells of the phloem, but from any place where the organic compounds are being loaded into the sieve tubes. Pressure is highest in the mesophylls, the photosynthetic tissue of leaves. The pressure is lowest at a sink, any region of the plant where the organic compounds are being unloaded or stored. That's what's known as the *pressure flow theory*.

Bio Buzz

Pressure flow theory states that internal pressure in the plant builds up at the source of the sieve tube system and then pushes the nutrient-rich solution on toward a sink, where the nutrients are unloaded.

Sucrose moves from the photosynthetic cells into the small veins in the leaves of the plant. From there, companion cells load the sucrose into the sieve tubes. As the sucrose concentration increases in the sieve tubes, water also starts to move in by osmosis—the movement of water across selectively permeable membranes, into different regions of the plant. As more water moves in, the pressure increases until, finally, the pressure pushes the sucrose/water solution out of the leaf, into the stem, and on to other parts of the plant—specifically, sinks, any region that requires nutrients.

Through the evolution of complex transport systems, vascular plants have mastered the art of staying alive under the most adverse of conditions.

In this chapter, we've mostly talked about how plants stay alive: what nutrients they need, how they obtain them, and how they transport these nutrients (and water) from the deepest roots to the topmost branches. In the next chapter, we'll be discussing how plants reproduce.

The Least You Need to Know

➤ Plants are autotrophs. Although they get most of their energy from sunlight, terrestrial plants still require 16 macronutrients and micronutrients for growth. These nutrients are obtained from the soil.

➤ The primary particle components of soil are sand, silt, and clay. The right mix of these three produces loam, the optimum growing medium for plants. In the right mix, sand, silt, and clay provide proper aeration for plant roots, as well as vital minerals.

➤ Vascular plants take in enormous amounts of water every day. The water and minerals are absorbed through the root system and are transported throughout the plant by xylem, a tubelike structure of cells that are dead at maturity.

➤ Transpiration, the movement of water from the root system out through the leaves, is accomplished through the power of evaporation. The drying power of air pulls the water upward in a fluid column.

Plant Reproduction and Development

With flowering plants found in nearly every corner of the globe, they've got to be doing something right. Besides having excellent systems for gathering and transporting water, minerals, and nutrients, plants are also excellent at reproducing. Mosses, ferns, conifers, and flowering plants all reproduce with incredible diversity.

In this chapter, we'll talk about flowering plants and how they reproduce. Despite the diversity found within the plant kingdom, there are many similarities when it comes to "baby" plants.

Reproductive Modes

Believe it or not, plants engage in sex, and they do so on a regular basis. Like humans, they produce sperm and eggs, and they have complex reproductive systems that house developing embryos. You and that daisy in your backyard have more in common than you think.

Although plants that have seeds have a reproductive advantage (in that they can wait out poor conditions and germinate when conditions are right), all plants have developed effective means of reproduction. During certain times of the year, for instance, moss gameotophytes produce reproductive structures on their tips.

All plants have male and female structures. The *antheridium*, the male structure, produces hundreds of sperm (all with flagella). The other structure, the *archegonium*, is the female structure, which produces only a single egg. Both produce the gametes by mitosis (which, if you recall, is done through cell division).

Plants reproduce in many ways. Before we move on to flowering plants, let's take a brief look at how a nonflowering vascular plant reproduces. For our purposes, we'll look at how mosses reproduce.

During wet periods, moss sperm break out of the antheridium and swim to an archegonium. Once there, they enter the antheridium's long neck, travel down to the egg, and fertilize it. The resultant zygote develops into an embryo. Through repeated mitotic division, the embryo forms into a multicellular sporophyte.

At first, the sporophyte is only a stalk, and it remains attached to the gametophyte for nourishment. But gradually, the cells at the tip of the stalk divide and form a *sporangium*, which in this case is called a capsule. The cells in the capsule undergo meiosis, turning into spores. The spores mature, the capsule splits open, and the spores are released.

The spores travel by water, currents of wind, and similar means. If the moss spore lands in a favorable environment, it will produce a slender thread called a protonema. The protonema is the earliest stage of moss gametophyte development. As it grows, it becomes a mature moss plant. In both mosses and ferns, water is the key ingredient to successful reproduction.

The reproductive cycle of conifers, another type of nonflowering plant, is a bit different. Conifers produce spores, but conifers and all other seed plants produce two types of spores. This phenomenon is called *heterospory*. The reproductive cycle of conifers is more complex than that of mosses and ferns, but the sequence is still the same.

This brings us up to the flashy members of the plant kingdom: flowering seed plants.

Bio Buzz

The male sexual organs of plants are collectively referred to as **antheridia**. The female sexual organs of plants are collectively referred to as **archegonia**.

Bio Buzz

Conifers and seed plants produce two types of spores, which is referred to as **heterospory**.

Hey, Sailor! Flowers!

Flowers exist for one reason only: to attract pollinators and ultimately allow the plant to engage in sex. These are the "red light district" of the plant. With their glorious colors, beautiful shapes, and enticing aromas, flowers are indeed worthy of our admiration and contemplation. They're lovely. But each flower is actually a set of unique, complex adaptations designed to assure the plant's reproductive success. The very traits that appeal to humans also appeal to insects, birds, and other pollinators.

Sexual reproduction is the primary reproductive mode for flowering plants. Many species of flowering plants also can reproduce asexually, but sexual reproduction is the norm. Mosses and ferns produce spores, which then produce new plants (eventually). Flowering plants work a little differently. In flowering plants, the entire plant is a *sporophyte*, or a body that grows through mitotic cells divisions from a fertilized egg (a bit like yourself).

At some point in the life cycle of a sporophyte, it will bear flowers, which are floral shoots that are specialized for reproduction. Within those flowers, both male and female gametophytes will form. Eggs will form within the female gametophytes, and sperm will form inside the male gametophytes. At fertilization, a cell will form that will contain genes from both gametophytes.

Flower Structure

Within the field of botany, there are botanists who do nothing but study flowers and their structure, which shows how specialized this part of a plant is. Flowers develop at the ends of floral shoots, which develop directly off the primary plant body.

Flowers are beautiful, but they're actually just modified leaves. Leaves usually are formed in four successive whorls; these modified leaves generally emerge from the end of a branch. As flowers form, they differentiate into nonfertile and fertile parts.

The fertile parts of flowers are called *stamens* and *carpels*. Stamens are the male reproductive structures of flowering plants, and carpels are the female reproductive structures. The two remaining, innermost whorls on a developing flower are composed of the plant's reproductive organs.

From Microspores to Pollen Grains

Each stamen has two parts. The *anther* produces pollen grains. The *filament* is the tall, narrow stalk that supports the anther.

Weird Science

Seed plants are the most successful plants on the planet. A clue to their success can be found in their reproductive cycle. In seed plants, such as conifers, the release of sperm and fertilization of the egg isn't dependent upon water. The sperm don't have to swim to the egg, and fertilization can take place during dry times.

Bio Buzz

A **sporophyte** is a vegetative body that grows through mitotic cell divisions from a fertilized egg. A cherry tree is a sporophyte, as is a delphinium.

269

The production of a male gametophyte, or pollen grains, is a complex process that takes place in the anther. In flowers, each anther contains four pollen sacs that contain "mother" cells. Through a process involving both meiosis and mitosis, each of these cells produces a two-celled structure, otherwise known as a pollen grain.

Of the two cells that make up the pollen grain, the larger one is the tube cell. This is the cell that will form the pollen tube into the egg. The smaller cell is the generative cell, which will form two sperm.

From Megaspores to Eggs

Carpels are the female reproductive structures of flowering plants. In most flowers, the carpels fuse, creating a structure called the pistil. Lots of flowering plant species have several stamens (the reproductive structures that produce pollen), sometimes as many as five, but most have only one pistil.

The base of the pistil (the fused carpels) forms the plant's ovary. This fusion is easy to see if you make a cross-section of any flower's ovary. The ovary usually is composed of chambers, each representing a folded carpel. Arising from the ovary, the fused carpels form into a stalk, called the *style*. At the tip of each style is a *stigma*. In most species of flowering plants, the stigma is sticky or is covered with hairs, both of which help trap pollen grains.

Flowers are composed of nonessential parts, such as sepals and petals, and essential reproductive organs, such as stamens and carpels, or pistils.

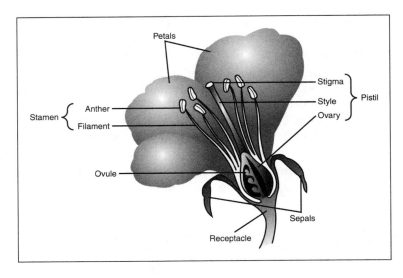

The ovaries are where the plant's *ovules*, or eggs, form. Ovules are the plant's megasporangia. As with their male counterparts, the production of eggs is a complex process involving both meiosis and mitosis. In the end, the ovule will have formed at least one, and frequently two, protective layers called *integuments,* which nearly encase the entire ovule. Only a small opening at one end of the ovule is left. Called the *micropyle*, this is where the pollen grains enter. Inside the ovule, the eight haploid cells (which is what all the meiosis and mitosis produces) arrange themselves into a structure referred to as the *embryo sac.*

Pollination and Fertilization

Every spring, flowering plants release billions of grains of pollen into the air. If you're a person who suffers from pollen allergies, you know exactly when this is occurring.

Pollination is the term used to refer to the transfer of pollen grains to a receptive stigma. Essentially, it's plants having sex. Pollination occurs in any number of ways. Wind is a well-known pollinator, but successful wind pollination depends on several factors. The plants need to release huge amounts of pollen, all at the same time. There has to be a fair amount of wind when this happens, or the pollen grains won't go anywhere but straight down to the ground. The individuals of the species also need to be located pretty close to each other—wind can carry lots of things a fair distance, but plants can't always count on having enough wind.

Although wind is effective, it offers a hit-or-miss method of pollination. Plants that have put a lot of energy into developing flowers do so to get a reproductive advantage. Those flowers attract insects, birds, and animals, which do the pollinating for the plant.

During pollination, pollen grains are inactive. But once a pollen grain lands on a receptive stigma (however it managed to get there), it becomes active again. It germinates. When pollen grains germinate, the tube cell kicks into action and starts forming the pollen tube. The pollen tube penetrates the stigma and grows down through the style to the ovule.

Meanwhile, the generative cell is dividing mitotically and is forming two sperm. The pollen tube grows through the micropyle and into the embryo sac, creating a passageway for the sperm. Both sperm travel through the pollen tube and enter the embryo sac of the ovule. One sperm fuses with the egg and forms a diploid zygote. The other sperm fuses with the remaining nuclei and, through repeated division, forms the *endosperm*, a nutritive tissue. *Double fertilization* is unique to flowering plants and really enhances their survival because each embryo comes with its own food supply.

Embryo Development

When we talk about the embryo development of plants, we're really talking about something most of us are very familiar with: seeds and fruit. A seed is a fertilized ovule, and ovules are found inside ovaries. A mature ovary is called a fruit. Upon fertilization, the new embryo embarks on a series of mitotic cell divisions. The result of all these cellular divisions is the development of seeds and fruit.

Recall that angiosperms (flowering plants) are classed into two categories: monocots and dicots. Recall also that one of the distinctions between the two is the number of cotyledons—seed leaves—present during embryo development.

Bio Buzz

Double fertilization is a process unique to angiosperms and involves two sperm. One sperm fertilizes the ovule and forms a zygote. The other sperm fuses with nuclei in the ovule, creating a food supply for the developing embryo called the endosperm.

Seeds

Let's look at the embryo development of two different seeds: a bean seed, which is a dicot, and a corn grain, which is a monocot. The monocot bean seed is kidney-shaped and is enclosed by a *seed coat*. The seed coat develops out of the ovule wall and the integuments of the ovule. Along one side of the seed is a small scar, the *hilum*, that marks the spot where the seed was joined to the ovary wall. Close to the hilum is the micropyle, the opening in the ovary where the pollen tube grew into the embryo sac. All these things are externally visible.

The majority of the bean's interior is composed of two large, fleshy cotyledons clamped together, almost like a clam shell. In beans, the cotyledons are the site of food storage.

In between the cotyledons is the rest of the embryo. The radicle is the embryonic root. The hypocotyl is a stemlike area between the radicle and the cotyledons. Last, but not least, is the epicotyl, a region at the top of the cotyledons. It and any embryonic leaves it contains make up the plumule.

Think About It

You may eat more fruit than you think. Using the true definition of "fruit," a lot of foods that we commonly class as "vegetables" aren't actually vegetables. Things such as green peppers, tomatoes, cucumbers, and okra are really fruits because they contain seeds.

Now let's look at a grain of corn, a monocot seed. A grain of corn is actually a fruit with a very thin wall. Corn is technically a fruit, but the seed occupies almost the entire grain. In corn, the cotyledon stretches over the radicle, hypocotyl, epicotyl (and corresponding plumule) like an umbrella, separating the embryo from the endosperm. In monocots, the cotyledon doesn't store food the way dicot cotyledons do. Instead, its function is to absorb nutrients from the starchy endosperm and transfer them to the developing embryo.

Fruit Formation and Seed Dispersal

Fruit forms around seeds, protecting them, aiding them in dispersal, and frequently providing a source of nourishment for the young seedling. Scientists have classed fruit into three major categories: simple fruit, aggregate fruit, and multiple fruit. Within each category are plenty of variations, but those are the three main groupings for fruit.

Simple fruits derive from a single carpel, or several carpels that have united. Essentially, simple fruits come from one simple or compound ovary. Nuts, grains, legumes, and achenes are all considered to be simple dry fruits. Hazelnuts, wheat, corn, rice, beans, peas, strawberries, and sunflowers also are simple dry fruits. Some simple fleshy fruits include oranges, lemons, olives, plums, peaches, grapes, and tomatoes.

Aggregate fruits derive from flowers that have several separate carpels. They come from separate ovaries, but all in the same flower. Raspberries and blackberries are an example of aggregate fruit. Multiple fruits derive from multiple flowers; they're a combination of a cluster of flowers. Pineapple and figs are multiple fruits.

There's one other category of fruit, called accessory fruits. These fruits are very similar to a simple fruit, in that they are derived from a simple ovary. But they differ from simple fruits because they come from a simple ovary *plus* other tissues. Bananas, apples, pears, melons, and squashes are all considered to be accessory fruit.

All these seeds and fruit are fine, but they don't do much for the continuance of the species if they just sit there. Nor do they do much if they just fall to the ground and land under the parent plant— the parent plant is bigger, stronger, and more developed, so it will garner all the sunlight and nutrients. The seeds need to disperse.

The same forces that aid in pollination—wind, water, and animals—also help with the dispersal of seeds. The fruits and seeds of plants that grow near water often contain air chambers that allow them to float. Coconuts, for instance, have so much air in the layers surrounding the seed that they can

Try It Yourself

You can easily observe the seed coat and cotyledons in a kidney bean. Soak a dried bean overnight in water. In the morning, the seed coat will be wrinkled, and you should be able to rub it off (often in one piece) with your fingers. With the seed coat removed, the bean will naturally split into two lengthwise halves. Each half is a cotyledon.

Weird Science

Seeds that disperse by way of air currents come with all sorts of adaptations. Orchid seeds are so tiny that they're dispersed by even the lightest gust of wind. Maple seeds have samaras, winglike appendages that cause them to spin and flutter through the air.

float in the ocean for hundreds or even thousands of kilometers without any damage. When they finally reach land, they germinate and produce a new palm tree. Until then, the entire time the seed is floating in the ocean; it's dormant.

Try It Yourself

When seeds germinate, how do they know which way to grow? You can see the effects of gravity on root germination through this simple experiment. You'll need a clear jar with a lid, some wet paper towels, and a large bean seed.

Soak your bean seed overnight in some water. (This isn't critical, but it will make everything occur faster.) The next day, place the bean seed inside the jar, positioning it on the side directly next to the glass. Then stuff the jar with fairly damp paper towels (which will hold the bean in place and provide moisture for germination). Then screw the lid on the jar, lay the jar on its side, and wait. Don't move the jar.

Within a few days, your seed should start germinating. When you've got about a quarter-inch of growth, turn the jar in the exact opposite direction, and wait. After a few days, check the jar. What has happened? Here's a hint: Your seed should still be growing and germinating, but the growth should now be crooked: germination is progressing, but it's responding to gravity.

Another way seeds disperse is by traveling on air currents. Unlike pollination, with seed dispersal it's not so critical that there be a strong wind or a simultaneous event such as the release of all the pollen grains at once. Because seeds can lie dormant without any ill effects, the haphazard dispersal methods of wind don't pose a huge problem.

Animals are great dispersing agents. Seeds can adhere to fur, feathers, or feet by way of sticky surfaces, spines, hooks, and hairs. Whither the animal goest, the seed goest. When animals eat fruit, they don't digest it all. Frequently, the seeds pass through the animal's digestive tract unharmed. If they end up in a place with favorable conditions, they'll germinate.

Humans may be the greatest dispersers of all. In the past, explorers carried seeds all over the world. Nowadays, the export and import of seeds is controlled in most places of the world, but seeds are already everywhere.

Growth and Development Patterns

The primary purpose of all this pollination, fertilization, and seed dispersal is to produce new plants so that the species can continue. When a seed lands on the ground, it doesn't just start sprouting. Plants have all sorts of mechanisms and adaptations to assure that the efforts that went into producing the seed and getting it to the brink of germination are not wasted.

Germination

A seed is part of a plant's life cycle, the dormant part. Different environmental signals will break the *dormancy* cycle and trigger germination. Water, temperature, light, and oxygen are the most common environmental stimuli for germination. Until the correct cues start filtering through, the seed will remain dormant, sometimes for hundreds or even thousands of years.

All seeds are dry. To germinate, they need to absorb water. Water softens the seed coat and triggers the release of enzymes that start turning the endosperm into simple sugars that the new embryo will use for its growth and development. Nearly all seeds need environmental temperatures of about 10° Celsius to germinate; with anything lower, they stay dormant. A germinating seed also needs oxygen for cellular respiration, and light plays a key role in seed germination as well.

Let's look at our dicot bean seed and see what happens during germination. The first visible sign of germination will be the emergence of the radicle. As soon as the radicle ruptures the seed coat, the shoot will begin growing. The hypocotyl will curve like a hook and break through the soil. As soon as it does this, it will straighten and pull the cotyledons and plumule up into the air. Embryonic leaves emerge from the plumule and begin to photosynthesize. As a result, the seedling begins to grow rapidly at the apical meristem, and everything really takes off. As soon as food moves into the growing plant—through photosynthesis and the root system—the cotyledons shrivel and fall off.

Weird Science

Cacao, corn, and oranges were all domesticated from seeds and were encouraged to reproduce in non-native habitats, and with great success. Other seeds have shown up uninvited. One of the symbols of the wide open western United States, the tumblin' tumbleweed, is an import. Prior to the arrival of Caucasians, it was unknown in the United States.

Bio Buzz

Dormancy is a state of decreased metabolic activity. Dormancy allows an organism to survive unfavorable conditions such as freezing cold or prolonged drought.

Our monocot corn grain germinates a bit differently. Unlike the cotyledons of the bean seed, the actual corn grain remains underground, transferring sugars from the endosperm to the developing shoots. The corn shoot isn't hook-shaped; it grows straight out of the grain, through the soil, and into the air. As it pushes through the soil, it's protected by a protective coat called a coleoptile. When the coleoptile breaks through the soil surface, the leaves of the plumule rapidly unfurl and photosynthesis starts.

Monocot and dicot seeds germinate differently. In monocots, the cotyledon stays underground, and the shoots emerge straight. In dicots, the cotyledons move above ground, and the shoots initially emerge in a hook-shaped pattern.

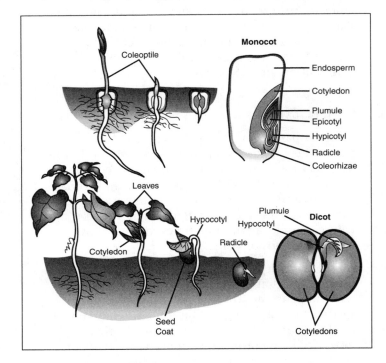

Weird Science

Some seeds require drastic conditions to germinate. The seeds of lodgepole pines, for instance, require a scorching fire before they can germinate. Other seeds, such as apples, will germinate only if they've been exposed to freezing temperatures first. Some desert plants have chemicals that inhibit germination. A deluge will wash away the chemicals, but a mere shower won't. This assures that the seed will germinate only when there is enough water available to produce adequate growth for survival.

We've covered a lot of material in this chapter, most of it complex. When it comes to reproduction, plants are no different than a lot of animals; they manufacture eggs and sperm, and then they develop ingenious ways to unite the two so that they can produce new plants. In the next chapter, we'll be talking about something different: the development and evidence for the theory of evolution.

The Least You Need to Know

➤ Sexual reproduction is the dominant mode of reproduction among flowering plants. Mosses, ferns, conifers, and angiosperms all reproduce sexually. Many plant species are able to reproduce asexually as well.

➤ Flowers develop at the ends of floral shoots. Sepals and petals comprise the nonessential nonreproductive parts. Stamens (males sexual organs) and carpels (female sexual organs) comprise the essential reproductive parts.

➤ In flowering plants, fertilization occurs when pollen grains are transferred to receptive carpel parts. A pollen tube grows out of the pollen grain, down to the ovule. Sperm travel down the pollen tube to the ovule. The resultant zygote becomes a seed.

➤ The growth and development of plants is accomplished through hormones, signaling molecules that work between cells.

Part 5

Nature: Always Moving Forward

"You've come a long way, baby," and guess what ... you're going even further. Species are always moving forward. From the first one-celled organism to the most complex human, life is never stagnant, it moves. Individuals obviously move, but so do species as a whole (which is one reason you look nothing like a Neanderthal or an ape—lucky you).

From Darwin's first speculations on evolution, to the development of genetic engineering, in the last 150 years, science has advanced more rapidly than any species ever did. And in the last 50 years, we've discovered more about human inheritance and our development as a species than in all the previous centuries put together.

We've also discovered how to burn fossil fuels, thin the ozone, deforest the rainforest, push other species to the brink of extinction, and acidify our forests. This is particularly troubling because for all our science and technology, we are still earthbound, confined to the only known habitable planet in the universe. Where do we go from here?

All living and nonliving aspects of this planet are interconnected, a truism that humans have come to late. We are the most destructive species on the globe, and if we want to continue life as we know it, even advance along the evolutionary line, we'll have to revise our thinking and change our ways. Can we do it? Let's give it a try.

So, this is "air"...

Evolution

At the very beginning of this book, we discussed evolution as it relates to the origins of life on Earth and the fossil record it left behind. But there's a lot more to it than just those concepts. In a technical sense, evolution is the theory that species change over time, and that means that evolution continues today.

The theory that all life on Earth (and you've now seen just how diverse those life forms are) is descended from a single common ancestor is mind-boggling. Just how strong is the evidence in favor of the theory of evolution? Perhaps more importantly, how does it occur? In this chapter, we'll try to answer some of those questions.

Darwin and His Theory

Since ancient times, humans have been attempting to unravel the riddle of life. At first everything, was explained in religious terms. But gradually, evidence arose about the nature of life that couldn't be explained in purely religious terms.

By the early 1800s, scientists were aware of the fossil record (and its implications), and they were aware of homologous and vestigial structures. (If you recall, homologous structures are structures that are similar across species, such as forelimbs in humans and penguins. Vestigial structures are structures that seem to serve no purpose, such as human tailbones and pelvic bones in snakes.) Many scientists began to suspect that some sort of evolution had given rise to the organisms populating the Earth. But there was no unifying theory to explain how evolution had come about. That it had occurred was beginning to be accepted, but *how* it occurred was a mystery.

In 1809, a French scientist, Jean Baptiste de Lamarck, presented the first systematic description of the evolutionary process, which he called "the inheritance of acquired characteristics." Simply stated, if an animal found itself in an environment in which survival was contingent upon traits the animal didn't have, the environment would cause the animal to develop these traits and pass them on to its offspring. Environmental pressures and internal "needs" could bring about permanent changes.

Lamarck's theory assumed the presence of "fluida." The driving force for change in an organism was centered in nerves that directed an unknown fluida to change body parts as needed. Based on Lamarck's theory, giraffes were originally short-necked, but finding themselves short of food, fluida directed their necks to permanently lengthen so that they could eat leaves previously out of reach. These long necks were then passed on to their offspring. "Fluida" is not an actual substance, it was Lamarck's own name for this unknown driving force.

Lamarck presented his theory but offered no supporting evidence. Consequently, his theory, while interesting, quickly fell out of favor within the scientific community, and no evidence in favor of it was ever found. But Lamarck hit upon one key point of modern evolutionary theory: The environment *is* an important factor in the evolutionary development of any species.

The next development in evolutionary theory came from an Englishman, Charles Darwin. Like many people of great genius, Darwin did not initially appear to have any talent, let alone extraordinary talent. From a very early age, he hated school and preferred to spend his time outdoors observing animals, birds, and insects. Nevertheless, at the age of 16, he was packed off to medical school to become a doctor. Unfortunately, the crude and painful procedures inflicted on patients sickened him. Losing patience, his father urged him to become a member of the clergy. So, Darwin went off to Cambridge to study theology.

At Cambridge, Darwin had good enough grades to earn him a degree in theology, but he spent most of his time with professors and faculty members who specialized in natural history. One botanist in particular, John Henslow, recognized Darwin's talent and encouraged it.

In 1931, at the height of evolutionary arguments within the scientific community, Henslow arranged for Darwin to be appointed as naturalist aboard the H.M.S. *Beagle*, a scientific ship about to embark on a five-year voyage around the world. Although the ship's main purpose was to more completely map coastlines, Darwin went along to gather scientific specimens. The young clergyman who hated studying suddenly found himself with the job of his dreams.

During the *Beagle*'s voyage, the captain frequently dropped Darwin off in one port and picked him up months later at another port. Darwin tramped hundreds of miles through unmapped regions, collecting hundreds of specimens and fossils. He recorded observations on thousands of different organisms. Back on board the ship, he would catalogue the specimens and write up his notes.

Ironically, one reason Darwin spent so much time on land was that he was constantly seasick. At every opportunity, he wanted off the *Beagle*, regardless of the climate or terrain. Without seasickness, he might never have gathered the number of specimens he did from the variety of locations that he did. The sheer volume and diversity of specimens became the cornerstones of Darwin's theory of evolution.

When Darwin returned to England in 1836, the scientific community universally praised his collections. Not only was the diversity astonishing, but Darwin had been impartial in all his observations and meticulous about his notes. As scientists began to study the specimens, some curious points emerged.

In studying some birds Darwin had collected from the Galapagos Islands, a small archipelago 1,000 kilometers west of South America, an ornithologist reported that Darwin had managed to collect 13 very similar but distinctly different species of finches. Each species of finch had a different beak. In addition, many of the fossils Darwin brought back were of extinct mammals. Paleontologists noted that several of these fossils were remarkably similar to present-day mammals.

These two points made Darwin think that perhaps the similarities between the finches meant that they shared a common ancestor. As for the similarities between the fossils and modern mammals, perhaps they indicated that an entire *species* could change over time, a process known as evolution.

In 1837, Darwin began working on his first notebook on evolutionary theory. He spent several years filling notebook after notebook with facts that could support his theory. He studied the fossil

Weird Science

A parting gift from Henslow, the book *Principles of Geology*, by Charles Lyell, greatly influenced Darwin because it spurred him to study landforms. During his five-year voyage, Darwin observed earthquakes, landslides, fossil shells 4,300 meters above the ocean, and more. Everywhere he went, he noted that all these geological events continually altered landscapes and that animals had to adapt to the changes.

Bio Buzz

A **species** is a group of organisms that look alike and that also are capable of producing fertile offspring in the natural environment. Organisms that just look alike may not necessarily be of the same species.

record extensively and intensively, noting that fossils of similar relative ages were more similar than fossils more widely spaced in age. He compared homologous, vestigial, and embryonic structures. He consulted animal and plant breeders of domestic species, conducting his own experiments. He also studied seed dispersal.

Through domestic breeding experiments with pigeons, Darwin concluded that a species could change rapidly. By selectively choosing certain traits, such as feathers on feet, or only white head feathers, and then breeding only birds with those traits, a pigeon species could change in only a few generations.

By 1838, most scientists accepted the notion that evolution occurred. But the real question was *how* it occurred. For Darwin, this was the burning question. Just as Henslow's parting gift of a geology book had caused him to view things differently, the clue to *how* also emerged from a book, this one called *Essay on the Principle of Population*, by British economist Thomas Malthus.

Malthus stated that an unchecked human population would double every 25 years, but that resources couldn't possibly increase at that rate. Consequently, humans eventually would find themselves in intense struggles for survival, competing fiercely for limited resources. It was this idea—the competition for limited resources—that gave Darwin the major piece to his puzzle.

Try It Yourself

The next time you're working on a new idea, remember that scientific advancement rarely works in a vacuum. Darwin's theory of evolution is a perfect example. Throughout his career, Darwin was always quick to acknowledge his sources, noting how other scientists provided him with clues, such as Lyall's book on geological changes and Malthus's book on competition among populations. "Genius" may boil down to persistence and knowing how to put the pieces together.

By drawing on the concept of competition, Darwin was able to explain how evolution could occur. First, he noted that variation existed among individuals of any species. Second, he stated that a burgeoning population would create a scarcity of resources in any given habitat. This scarcity of resources would create competition among the individuals of a species: competition for sheer survival. Such a competition would lead to the death of some individuals and the survival of others.

Based on these premises, Darwin concluded that the individuals with advantageous variations were the ones that would survive and would reproduce more individuals with those same advantageous variations.

Natural Selection

Darwin coined the term *natural selection* to describe the process by which individuals with advantageous variations survived and reproduced other individuals with the same advantageous variations.

As an example of natural selection, Darwin pointed out ptarmigans that turn white in winter. (Ptarmigans are wild, chickenlike birds that have white feathers in winter and mottled brown feathers in summer.) By inference, he noted that this color change allowed ptarmigans to survive in winter. Ptarmigans that didn't change color in winter were easily spotted by predators and then promptly eaten. Therefore, Darwin speculated, ptarmigans that possessed the trait of changing color were more likely to survive and pass on this advantageous trait to their offspring.

Bio Buzz

Natural selection is the process by which organisms with favorable variations survive in the natural environment and reproduce more individuals at higher rates than those organisms without the advantageous variations.

Darwin was no fool. He anticipated that his theory of evolution, and how it occurred, would cause controversy. Consequently, he waited to announce it, compiling notebook upon notebook of evidence, looking for flaws in his reasoning. He waited so long that he might never have published his evidence. But in 1858, another naturalist, Alfred Wallace, came up with virtually the same theory.

While living on the Malay Archipelago in the Pacific Ocean, Wallace, using Darwin's research and data (which Darwin had made available to the entire scientific community), had independently come to the same conclusions as Darwin years before. Wallace formulated the key points into an essay and sent it off to Darwin and other scientists. Darwin's colleagues stepped in and prevailed upon him to let them announce the theory by formally presenting his notebooks and Wallace's essay at the same time. Both Wallace (who believed that all the credit should go to Darwin) and Darwin agreed.

The presentation itself excited very little reaction, but one year later, in 1859, Darwin published an extraordinary book, *The Origin of Species*, that changed the face of biology forever. The first printing of the book sold out in one day. In Darwin's own lifetime, most scientists came to accept the theory of evolution as true. Seventy years later, advances in the fledgling field of genetics proved him correct.

Bio Buzz

Adaptive radiation occurs when many related species evolve from a single ancestral species. When two or more related species become more dissimilar, the process is called **divergent evolution.** **Convergent evolution** takes place when two unrelated species become more similar in appearance as they adapt to the same type of environment. **Coevolution** involves the joint change of two species in close interaction with each other.

Weird Science

Some animals, such as bats, actually coevolved with flowering plants. Bats that feed on flower nectar have long, slender muzzles and long tongues with a brushed tip. When the bat feeds, pollen sticks to its facial fur. When it visits the next flower, the pollen is transferred. Flowers that coevolved with bats are lighter in color than other flowers, which helps the nocturnal bats find them.

After Darwin presented his theory, the main point of contention among his colleagues concerned what came to be known as the "missing links." If species did indeed evolve, where were the transitional forms, the species that bridged the evolutionary gaps between major groups of organisms? For instance, as a group of organisms, birds are unique. But they didn't just happen. They had to have somehow evolved from the primordial soup of all organisms. Where were the fossil forms that show the evolutionary jumps from general, nonspecialized organisms to specific groups of organisms, such as birds, snakes, or other organisms? Oddly enough, the evidence was already there, but people just didn't recognize it. It would take another century of additional fossil finds and advances in molecular biology to yield plausible answers to the "missing link" question.

A fossilized skeleton, about the size of a pigeon, had been unearthed in a limestone quarry in southern Germany. Originally labeled as a small meat-eating dinosaur, the fossil possessed clawed fingers, a heavy jaw with short, spiky teeth, and a long, bony tail. Later, another fossil of the same type was unearthed. Still later, someone noticed the feathers. Both specimens had feathers.

Feathers? What were feathers doing on a dinosaur? People thought it was a hoax, but it turned out to be true. *Archaeopteryx* was the missing link, the evolutionary bridge, the jump between species. It was the jump scientists were looking for, the jump in which a group of organisms went from land-dwelling to air-dwelling.

Nowadays, scientists recognize natural selection as the driving force of evolution. Natural selection pushes evolution into four patterns: *adaptive radiation, divergent evolution, convergent evolution,* and *coevolution.*

The finches of the Galapagos Islands, also known as Darwin's finches, are a good example of adaptive radiation. The finches on the Galapagos are presumed to have come from a common ancestor, a bird that managed to make it across the more than 900 kilometers of treacherous open water. Yet the environment has caused the finches to adaptively radiate into other species. One species has a beak adapted to breaking hard-shelled seeds, another has a beak adapted to small seeds, and yet another species is adapted to eating insects—the list goes on and on.

Red foxes and kit foxes are an example of divergent evolution. Both species are clearly related, yet they have undergone divergent evolution. The red fox lives in forests and grasslands, and its dark coat helps it blend into its environment and survive. The kit fox lives in deserts and on open plains, and its pale coat blends in with the surrounding grasses and sand. The ears of kit foxes are also larger than those of red foxes. The larger surface area enables kit foxes to get rid of excess body heat, which is an important trait in a hot desert environment.

Cacti offer a good example of convergent evolution. Entirely different species of cacti, all with similar adaptations, have evolved in different desert habitats around the world. The giant saguaro cactus of Arizona bears a remarkable similarity to the spurge *Euphorbia*, which is found in the deserts of Africa. Both have developed long fleshy arms with spines and a thick epidermal covering.

We've already looked at the coevolution of two major groups: flowering plants and insects. Each group evolved in conjunction with the other, and each evolved more rapidly because of the other's presence.

The Origins of Humans

The Origin of Species was a bestseller that caused a fair amount of controversy. Twelve years later, Darwin published a second book, *The Descent of Man*, another treatise considered radical for its time. In it, he cited evidence that humans were most closely related to a group of mammals known as *primates*. Darwin's work remains a landmark in the scientific study of human origins.

Paleoanthropologists, scientists who study human evolution, face a difficult task. Most of our knowledge of human evolution has been acquired through the study of fossilized remains. But needless to say, humans are quite adaptable, and they were very mobile even in primitive times. By the very nature of their lifestyle and habits, the human fossil record is spotty, at best.

Thus, paleoanthropologists must make deductions and inferences as to the origins of humans based on very subtle clues. For instance, from a fossilized piece of curved skull, they can deduce the cranial capacity, or size of the brain case. By studying the wear and tear of a tooth, they can infer the diet and nutritional habits of ancient man. The markings on fossilized limb bones enable scientists to ascertain what muscles were attached, how large they were, and the mode of locomotion. Radiocarbon dating allows paleoanthropologists to place each fossil correctly on a time line. Because human fossils are few and are rarely complete, each new find offers significant amounts of information.

To even begin to classify human fossils, scientists must first determine how humans are related to other animals. Basing their conclusion on fossil evidence, comparative

Bio Buzz

A **primate** is a specific class of mammals. Most primates have highly movable fingers and toes that have flattened nails instead of claws, good vision, a larger cranial capacity, and the ability to hold their bodies upright.

Bio Buzz

An **anthropoid** is a primate that is characterized by the presence of an opposable thumb and a large cranial capacity relative to body size.

morphology, and biochemistry, scientists now know that humans belong to the primate class of mammals.

When mammals first appeared on Earth, not all of them lived exclusively on the ground; some of them moved into trees and lived there. The first primates evolved from these tree-dwelling mammals more than 60 million years ago. The fact that they lived in trees helps explain some of their features today: They needed fingers and toes to grasp trees and, good vision allowed them to judge distances when swinging from branch to branch. An upright position allowed them to feed while still keeping an eye out for predators, and they also were able to flee instantly. Today, all primates, including humans, still retain these characteristics.

Monkeys, apes, tree shrews, tarsiers, and humans are all primates. Within the primate class, monkeys, apes, and humans form a subgroup called *anthropoids*. Anthropoids evolved from primates at least 36 million years ago.

In comparison to other primates, apes (including gorillas, chimpanzees, orangutans, and gibbons) and humans have a large cranial capacity, relative to body size. An opposable thumb—a thumb that can be positioned opposite of every finger—really sets anthropoids apart from other primates and gives them a precision grip. Humans and apes also have evolved to be capable of sitting, standing, and walking erect; unlike most other primates, they also have no tails. Within the anthropoid grouping, apes, humans, and extinct species of the human lineage are further subclassed as *hominoids*. Hominoids evolved from anthropoids somewhere between 10 million and 5 million years ago.

About five million years ago, there was a divergence from the last shared ancestor of apes and humans. Whatever caused this divergence—and no one knows what—humans made the jump into a completely separate grouping, unlike anything else on Earth. Hominids had appeared on Earth.

From Apes to Neanderthals to Humans

Hominids are a subgroup of primates that includes humans, known as *Homo sapiens*, and their immediate ancestors. Although *Homo sapiens* diverged from other primates about five million years, in the ensuing five million years, humans themselves have also evolved. Humans today are not the same as the first humans that appeared five million years ago.

As a group, hominids possess some unique traits that set them apart from other primates. For instance, bipedalism, the ability to walk upright on two legs, is a uniquely human trait. Occasionally other primates walk upright, but mostly they walk on all fours or swing from branches. The human pelvis is adapted for upright walking: It supports the internal organs when walking upright. The shape of the human foot is also uniquely adapted for walking upright. The bones of the big toe are aligned with the other toes, which causes the weight to distribute evenly.

Human cranial capacity also sets us apart from other primates. The high forehead of humans indicates this increased cranial capacity. This front part of the brain also contributes to another uniquely human trait: the ability to communicate through verbal language. Apes communicate through grunts and sounds, and many have learned to communicate through sign language. In the wild, however, apes have never developed the complex, sophisticated system of signals that characterizes human language.

Among all the organisms on Earth, humans appear to be the only ones that have developed and sustained a culture.

The first known genus of hominids showed up about five million years ago. During this period, a wide variety of hominids appeared on the scene. So far, scientists have had a hard time deciphering their family tree and simply group them as *australopiths*. The first australopithecine fossils were found in South Africa, but others have subsequently been found in eastern Africa.

Although it's hard to class the *australopiths*, there are some distinctions. The oldest australopithecine fossils are classed *A. afarensis*. ("A." is an abbreviation for *Australopithecus*.) They were shorter than present-day humans, only about 1.5 meters tall. Their cranial capacity was only about 380 to 450 cm³, less than a third of the capacity of today's humans.

Weird Science

Compared to other primates, the cranial capacity of humans is markedly larger. For example, chimpanzees have a cranial capacity of 500 cm³. But humans have a cranial capacity of 1400 cm³, nearly three times as large.

Weird Science

The transitional species that were so critical to the acceptance of Darwin's original theory can be found in most lineages, including humans. In the 1930s, two skulls were found, one in Great Britain and the other in Germany. Both were between 250,000 and 300,000 years old. Each skull had the distinct brow ridges of *H. erectus* and the larger cranial capacity of *H. sapiens*; thus, on the time line, these two specimens were transitional species between the two groups.

The next oldest *australopiths*, A. *africanus*, lived about 2.2 million to 3 million years ago, and was also first found in South Africa. A. *africanus* was taller and heavier than A. *afarensis* and had a cranial capacity between 430 and 550 cm^3. Two other australopithecine species, A. *robustus* and A. *boisei*, arose after A. *africanus*. Again, the trend was toward greater height and larger cranial capacity.

Two other examples of hominids, *Homo habilis* and *Homo erectus*, also appear on the hominid time line, each progressively taller and with larger cranial capacity. *H. habilis* was the first fossilized human that showed signs of working with tools and that had a brain that was developed enough for speech. *H. erectus* used stone tools, ate plants and meat, and most likely learned how to harness and use fire. The next evolutionary jump was to *Homo sapiens*, the first true ancestors of the humans we recognize today.

Early hominid fossils have been found primarily in Africa. But younger human fossils—those that range from 35,000 to 130,000 years old—have been found in Europe, Asia, and Africa. These fossils belong to an early group of *H. sapiens* called Neanderthals.

Neanderthals had heavy bones, very thick brow ridges, and small chins. They were about 5 feet tall and were stocky. Oddly enough, at least as far as our own evolutionary vanity is concerned, they had larger brains than we do; the cranial capacity of a Neanderthal was about 1450 cm^3. Neanderthals lived in caves and stone shelters, and were adapted to the colder climate of northern Europe.

About 35,000 years ago, yet another evolutionary adjustment occurred in humans. The cranial capacity decreased to 1400 cm^3 (the cranial capacity of today's humans), a high forehead and a prominent chin developed, and brow ridges disappeared. Cro-Magnon, the *H. sapiens* that scientists regard as the modern human, had evolved. ("*H.*" is an abbreviation for *Homo*) Cro-Magnons had a sophisticated culture: They hunted, fished, and gathered. They also made tools, including needles and fishhooks.

About 50 years ago, four young boys discovered an intricately tunneled cave in the Perigord region of France. Deep in the maze of tunnels was a chamber covered with magnificent sketches, engravings, and paintings. Because the chamber was cool, dry, and virtually sealed off from the outside environment, the red, brown, yellow, and purple pigments were as brilliant as the day they were painted. Radiocarbon dating revealed that the paintings were between 17,000 and 20,000 years old. Even earlier art, approximately 25,000 years old, has been found in the Pyrennees, a mountainous region of France.

Since the advent of Cro-Magnons, our direct ancestors, human evolution has been primarily cultural rather than biological.

Lucy and the Black Skull: The Missing Links?

Paleoanthropology is fraught with controversy, perhaps even more so than other branches. The fossilized specimens are few and far between, and although radiocarbon dating can accurately place them on a time line, nearly everything else is subject to wide interpretation. Take the cases of Lucy and the Black Skull, for instance.

In 1974, paleontologists unearthed the oldest and most complete fossilized skeleton in eastern Africa. The shape of the fossil's pelvic bones indicates that the remains are female. "Lucy," as its discoverer dubbed the australopithecine fossil, lived about three million to three and a half million years ago. Lucy is classified as an *A. afarensis*.

In 1985, an unusual fossilized skull was found in Kenya. Richard Leakey and Alan Walker, the paleoanthropologists who found it, dubbed it the Black Skull because of its unusual color. The hominid who was the Black Skull lived about two and a half million years ago, and Leakey and Walker have asserted that it is of the species *A. boisei*.

Not so, say other paleoanthropologists, including Donald Johanson, the discoverer of Lucy. Johanson and his colleagues compared 32 characteristics of the Black Skull with known skeletons of *A. boisei*, its contemporary, *A. robustus*, as well as *A. afarensis*. (*A. robustus*, *A. afarensis*, and *A. boisei* are considered "contemporaries" because they all existed on earth during roughly the same time period.) They found the Black Skull to contain a baffling mix of both primitive and sophisticated traits, only two of which are shared exclusively with *A. boisei*.

As a consequence, many paleoanthropologists believe the Black Skull to be a transitional species, past *A. afarensis*, but somewhere between *A. africanus*, *A. robustus*, and *A. boisei*. Leakey and Walker maintain that the Black Skull is an example of *A. boisei*, only with unusual features.

Who is right? Does the Black Skull point toward another lineage, or is this one fossil representative of a highly unusual individual? At present, no one knows.

What we do know is this: The traits that we think of as uniquely human didn't just evolve overnight or even over thousands of years. They probably didn't just evolve from one lineage, either. Primates in general, and *H. sapiens* in particular, are the result of a complex and sophisticated interweaving of lineages. If we ever successfully unravel the story of life on Earth, especially our own, it probably won't be in our lifetime.

We've covered some interesting and thought-provoking material in this chapter. Just where did we (or any species, for that matter) come from? From Darwin's first musings to modern discoveries of ancient fossils, the origin of humans is a puzzle with many missing pieces. In the next chapter, we'll discuss genetics, a possible key to evolutionary development.

The Least You Need to Know

➤ Evolution is the theory that species change over time.

➤ Darwin noted that within any species there are individual variations. The individuals most likely to survive that competition were those with advantageous variations. In turn, they would pass these advantageous traits on to their offspring.

➤ Darwin coined the term *natural selection* for the process by which individuals with advantageous variations survive in the natural environment and reproduce offspring with the same advantageous traits faster than those without the advantageous traits.

➤ It is difficult to ascertain human evolution with any certainty because human fossil remains are few and far between. Scientists believe that humans evolved from primates, the mammalian group that includes monkeys, apes, and humans. But about five million years ago, the human lineage diverged from its last shared primate ancestor. Modern humans have existed for approximately 35,000 years.

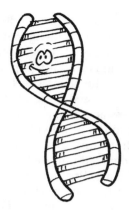

Genetics

According to Darwin, the natural environment was a harsh place. Only those individuals with advantageous variations would survive, reproduce, and pass on those advantageous variations to their offspring, who in turn, would also possess those advantageous traits and, theoretically, have a better chance of survival.

What are these advantageous traits, and how does an individual pass them on to its offspring? The "traits" that so intrigued Darwin are now represented in what we call genes. Although Darwin didn't have a name for it, his theory of evolution and natural selection was dependent upon genetics, the study of genes and the way they work. That's what we'll discuss in this chapter.

A Man and His Peas: Mendel

In the mid-1800s, people were beginning to seriously ponder the mechanics of evolution and *inheritance*. At that time, people accepted the idea of *heredity*, in that the sperm and eggs transferred certain traits to offspring, but the common belief was that the two sets of information blended together into a uniform blob at fertilization. People never considered the possibility that these traits were transferred in units.

However, if anyone had given this serious thought, the blending idea wouldn't ultimately work. If this was true, why weren't all the descendants of white mares and black stallions gray?

Bio Buzz

Inheritance is the process of passing on traits through heredity. **Heredity** is the transmission of traits from parents to their offspring.

Even with all the obvious flaws in the blending of inherited traits theory, few people challenged it. Charles Darwin did. Based on his theory of natural selection, Darwin asserted that individuals of any population exhibited variations. Through the generations, the advantageous variations were passed on to offspring, who also survived. The less advantageous variations persisted among fewer and fewer individuals, or even disappeared. Darwin didn't think that the less advantageous variations were blended out; he thought that they just appeared less frequently over time. As Darwin pondered the question in England, even before he ever presented his theory, someone else was pondering the same questions of inheritance in Austria: Gregor Mendel.

Gregor Mendel was a contemporary of Charles Darwin, in that each lived during the other's lifetime. Although it's unknown whether Darwin knew of Mendel's work, Mendel most likely knew of Darwin's work. Mendel was 37 when Darwin published *The Origin of Species*, and it's unlikely that the furor the book created escaped anyone, even an Augustine monk in Austria.

Mendel grew up surrounded by nature and was fascinated with plants. He came from a farming family and was fully aware of agricultural principles and their applications. In 1843, at the age of 21, Mendel entered the monastery of St. Thomas, destined to become an Augustine monk. Because of his farming background, Mendel was the logical choice to tend the monastery garden.

Although the monastery of St. Thomas was somewhat removed from the centers of scientific enquiry, Mendel kept abreast of all the latest agricultural developments and breeding experiments. He was a member of the regional agricultural society, and he won several awards for developing improved varieties of vegetables. He was not a man of narrow interests who accidentally stumbled onto a great notion; he scientifically pursued agricultural advancement.

Upon his return to St. Thomas, Mendel began a series of controlled experiments using plants. He wanted to determine whether a predictable pattern to inheritance existed. Mendel chose pea plants for his experiments because his family farmed them and because he was already familiar with them. Pea plants are self-pollinating, in that both male and female gametes form on different parts of the same flower. Successive generations of pea plants are just like their parents in one or more characteristics.

Mendel had observed that some tall pea plants produced short offspring, while others produced tall offspring. Some pea plants with yellow seeds produced offspring with yellow seeds, but others produced offspring with green seeds. What Mendel was observing was inheritance.

To see whether there was a predictable pattern to inheritance, Mendel decided to study seven characteristics of pea plants, each of which occurred as two contrasting traits:

1. Seed color (green or yellow)
2. Seed texture (smooth or wrinkled)
3. Seed coat color (white or colored)
4. Pod color (yellow or green)
5. Pod appearance (constricted or inflated)
6. Flower position on the stem (axial or terminal)
7. Stem length (short or tall)

By studying the occurrence of these traits, Mendel hoped to discover a predictable pattern of inheritance. To start, he spent two years developing *pure strains* by letting the plants self-pollinate. Plants that are *pure* for a particular trait always produce offspring with that trait. A *strain* is a term used to denote all plants that are pure for that trait.

Weird Science

Mendel believed that parents pass on traits to their offspring, but he did not know of the existence of DNA or genes. Nowadays, scientists believe that genes are comprised of segments of DNA molecules. A gene is the segment of DNA on a chromosome that controls a particular hereditary trait. Genes aren't particles, per se, but they act like particles would when transferred from generation to generation.

Think back to plants: Pollination is the transfer of pollen grains from anthers to a receptive stigma. In allowing the plants to self-pollinate, Mendel obtained 14 strains, each one pure for its trait (yellow seed color, short stem, and so on). He called each of these strains the parental generation, or the P1 generation, for short.

Mendel then proceeded to cross-pollinate the strains by transferring pollen from one plant to the stigma of a plant with a contrasting trait—for instance, he transferred pollen from a long-stemmed plant to the stigma of a short-stemmed plant. He then allowed the seeds to develop and recorded how many offspring of each type each plant produced. Mendel called these offspring of the P1 generation the first filial generation, or the F1 generation, for short.

He then self-pollinated the flowers from the F1 generation and collected the seeds that resulted. He called the plants from this generation the second filial generation, or the F2 generation, for short. By following this process, Mendel performed thousands of cross-pollinations, documenting the results each time.

In one of his first experiments, Mendel crossed two pure strains, one pure for green pods, and the other pure for yellow pods. To his surprise, the F1 generation produced only green-podded plants even though the other parent had been a pure strain for yellow pods. Of two traits found in the P1 generation, only one of them showed up.

If Mendel stopped experimenting at this point, he would have concluded that crossing green pod plants with yellow pod plants always produced green pod offspring.

However, when the next generation—the F2 generation—developed, three-fourths of the offspring had green pods, and one-fourth had yellow pods. The lost trait had reappeared. This particular pattern of a disappearing trait that re-emerged in a ratio of 3:1 showed up in thousands of cross-pollinations.

Three Mendelian Principles

After studying the results of thousands of cross-pollinations, Mendel concluded that inheritance was governed by three basic principles:

1. The Principle of Dominance and Recessiveness
2. The Principle of Segregation
3. The Principle of Independent Assortment

The Principle of Dominance and Recessiveness

Knowing nothing of chromosomes and genes, Mendel theorized that something in a plant was controlling the expression of a particular characteristic. He called this something a factor. Every characteristic Mendel studied had two traits—for instance, the seed could be either yellow or green. He theorized that each characteristic was the result of the interaction of a pair of factors.

He next looked at how the factors interacted by studying the F1 generation. In attempting to explain the appearance of only one trait, Mendel arrived at the *Principle of Dominance and Recessiveness.*

Mendel called one of the factors dominant because it was dominant over the other factor and masked its trait. He called the other factor recessive because it was masked in the presence of the dominant factor. He concluded that factors that produced visible traits in an F1 generation were dominant over the invisible recessive factors of that generation.

Weird Science

The principles that Mendel established in the 1860s are the foundation for the modern field of genetics. Because of this, this branch of science is referred to as Mendelian genetics. It is also called classical genetics, to distinguish it from molecular genetics, another branch of biology.

Bio Buzz

The **Principle of Dominance and Recessiveness** states that one factor in a pair may mask the other factor, preventing it from having any effect.

Try It Yourself

You can do a quick survey of dominance and recession right in your own family. Who's got brown eyes? That's a dominant trait. Who's got blue eyes? That's a recessive trait. Now you probably could have figured out that eye color was a genetic trait on your own, but what about tongue-rolling? Can anyone in your family roll their tongue? That's also a dominant genetic trait.

The Principle of Segregation

The next thing Mendel was curious about was why the recessive traits of an F1 generation would reappear in subsequent generations. He theorized that if each parent had two factors, then each offspring must also have two factors. Each parent couldn't be passing two factors, or each offspring would then have four.

Mendel concluded that each reproductive cell received only one factor for each trait. Although he knew nothing about meiosis (the process by which duplicated chromosome numbers reduce by half in preparation for the production of a new daughter cell), Mendel's *Principle of Segregation* perfectly describes the process.

Bio Buzz

The **Principle of Segregation** states that the two factors for a trait segregate during the formation of sperm and eggs.

The Principle of Independent Assortment

Mendel crossed more than pure strains. He also crossed plants that displayed two characteristics, such as long-stemmed plants that had yellow seeds. When he studying the data from these complex crosses, it indicated that dominant factors did not necessarily appear together. The dominant factor of a green pod could show up in a short-stemmed plant, which itself was produced by recessive traits.

Bio Buzz

The **Principle of Independent Assortment** states that factors for different traits are distributed independently in reproductive cells.

Mendel concluded that the factors for different traits were not connected. Based on this, he formulated his *Principle of Independent Assortment.*

Mendel performed his experiments during the 1860s and 1870s. In 1865, he presented a paper detailing his findings. But because no one knew about meiosis or the existence of chromosomes, there was no way to explain his findings. Also, at that time, no scientist had used statistics to explain biological phenomena. Mendel's pairing of mathematics and natural events was considered downright odd.

Mendel didn't live to see his principles accepted and embraced. His work remained sunk in obscurity until 1900. In that year, it was rediscovered three separate times by three different scientists working independently. Shortly afterward, Walter Sutton, a biologist working with genetics, proposed the theory of chromosomes, which linked Mendel's principles directly to what was known about meiosis. Nowadays, statistics and other quantitative methods are used extensively in any type of scientific research.

Instead of "factors," scientists study genes, the DNA segment on chromosomes that controls a particular hereditary trait. Because chromosomes occur in pairs, genes also occur in pairs. Each of the several alternative forms of a gene—the equivalents of Mendel's "factors"—are called alleles.

Subsequent scientific research has also supported Mendel's principle of segregation because we now know that when gametes are formed, they receive one chromosome from each homologous pair, and during fertilization, the offspring receives one allele from each parent. We also know that when chromosomes segregate independently, any alleles on them also segregate independently, which supports the principle of independent assortment.

Without ever seeing what was going on, Mendel was able to correctly ascertain inheritance patterns. Nearly 150 years after he stated them, his principles are still correct.

Think About It!

If you ever want to decipher a genetic code, it's pretty simple, particularly if you know the alphabet: Scientists use letters to represent alleles. A capital letter refers to a dominant allele, while a lowercase letter refers to a recessive allele. In pea plants, the dominant allele for tallness is represented by the capital letter T, while the allele for shortness is represented by the lowercase letter t.

Patterns of Inheritance

By coupling statistics with biological phenomena, Mendel had hit upon a brilliant pairing. Many fields of science lend themselves to statistical analysis, but none probably so well as genetics. Inheritance produces patterns, most of them predictable.

The genetic makeup of any organism is called its *genotype*, and the external appearance of any organism is called its *phenotype*. For instance, the genotype for one of Mendel's pure-strain tall pea plants would be TT because it consists of two dominant alleles for stem length. The same plant's phenotype, or appearance, would be "tall." In an F1 plant with one tall parent and one short parent, the genotype would be Tt (because each parent contributed an allele), but the phenotype would still be "tall" because the T is dominant over the t.

If both alleles are the same, as is the case of the pure-strain tall plant (TT), then the organism is said to be homozygous for that particular trait. Organisms can be homozygous dominant or homozygous recessive. A tall pure-strain plant (TT) is homozygous dominant. A short pure-strain plant (tt) is homozygous recessive.

If the two alleles of a pair are different, the organism is said to be heterozygous for that trait. The F1 generation of a cross between TT and tt will be heterozygous because the genotype of each plant will be Tt—the alleles in the pair are different. Yet, the phenotypes of all the F1 plants will be "tall."

Bio Buzz

The **genotype** of an organism is its genetic makeup. The **phenotype** of an organism is its physical appearance.

Mendel's principles worked as flawlessly as they did because he studied only traits that had two alleles. Scientists now know that some traits are determined by multiple alleles. Human blood is a good example of a trait controlled by multiple alleles. The alleles for human blood are referred to as IA, IB, and iO. From these three there are several combinations possible, yet each individual can inherit only two alleles for any given trait, one from each parent.

Much of the study of genetics is based on *probability*, the likelihood that a specific event will occur.

Probability = the number of one kind of event ÷ the number of all events

Probability is usually expressed as a percentage, a decimal, or a fraction. For instance, in Mendel's experiments for the dominant trait of yellow seed color over the recessive trait of green seed color, yellow seeds appeared 6,022 times in the F2 generation. The total number of individuals was 8,023—6,022 divided by 8,023 equals 0.75. So, the probability for the occurrence of yellow seeds is 75 percent, or three-fourths, or 0.75.

Geneticists use a Punnett square (so named for its inventor, R.C. Punnett) to help them predict probabilities. Punnett squares are useful for probabilities because they can be expanded to suit any number of alleles.

Several types of monohybrid crosses can occur: homozygous × homozygous, homozygous × heterozygous, heterozygous × heterozygous, testcross, and codominance.

A testcross allows you to determine the unknown genotype of a particular phenotype. For instance, if you have a black rabbit, you don't know whether its genotype is BB or Bb because both genotypes will produce the phenotype "black." In a testcross, an unknown genotype is crossed to a known homozygous recessive genotype. The results of a testcross, graphed on a Punnett square, may be able to reveal the unknown genotype.

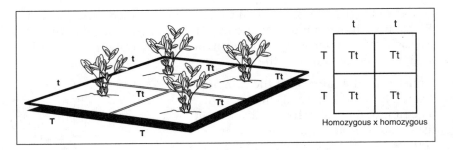

Punnett squares, invented by R.C. Punnett, are a method of graphing probabilities. These squares show the probabilities of two types of monohybrid crosses: homozygous × homozygous, and homozygous × heterozygous.

Bio Buzz

Codominance, or incomplete dominance, occurs when two or more alleles influence the phenotype of an organism.

Bio Buzz

A group of genes located on a chromosome is called a **linkage group.** Genes in a linkage group usually are inherited together.

Mendel studied traits in which one allele is dominant over the other. But not all alleles possess dominant or recessive traits; some alleles are codominant. *Codominance* results in an intermediate trait, something between dominant and recessive. The flower called Japanese four o'clocks is a good example of codominance. Four o'clocks produce both red and white flowers, but neither trait is dominant. When red-flowered plants self-pollinate, they produce red flowers. When white-flowered plants self-pollinate, they produce white flowers. But when a red and a white plant cross-pollinate, they produce pink offspring. What would happen if you cross-pollinated two pink-flowered four o'clocks? Look at the Punnett square.

Human blood types are also an example of codominance because no one allele is dominant over another. If you have two A alleles, your blood type will be A. If you have two B alleles, your blood type will be B. But if you have an A and a B, allele, your blood type will be AB. If you have blood type O, it means that your blood type is neither A nor B.

In addition to predicting the outcome of crosses through Punnett squares, some genetic outcomes can be predicted through *linkage groups*. Genes are found on chromosomes. Because there are a limited amount of chromosomes (23 pairs in humans) and thousands of traits, there obviously can't be a chromosome for every trait. Each chromosome must carry many genes.

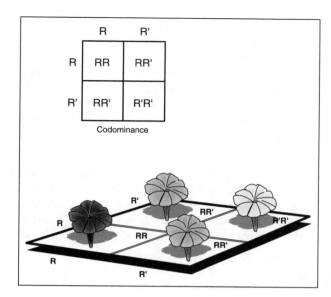

Codominance

Cross-pollinating two pink-flowered four o'clocks will produce two pink-flowered offspring, one red-flowered offspring, and one white-flowered offspring.

In the early 1900s, a geneticist at Columbia University named Thomas Hunt Morgan began breeding experiments using the common fruit fly, *Drosophila melanogaster*. Fruit flies are ideal subjects for breeding experiments because they have a generation time of only 10 to 15 days. Each mating produces hundreds of individuals, the flies have easily distinguishable physical traits, and they are easily kept in a laboratory. Fruit flies reproduce so well, in fact, that Morgan's laboratory at Columbia quickly became known as "the fly room."

In studying *Drosophila*, Morgan observed that in the females, one pair of chromosomes was identical. In males, one chromosome looked like that found in the females, but the other was shorter and hook-shaped. He called the large female chromosomes the X chromosomes, and he called the shorter male chromosome the Y chromosome.

Morgan was the first scientist to correctly hypothesize that the X and Y chromosomes are sex chromosomes, or chromosomes that determine an individual's sex. All other chromosomes in the body—the ones not involved in sex determination—are called autosomes.

This discovery led Morgan to hypothesize that some traits were always linked to one sex or the other, and his experiments confirmed this. One of his experiments confirmed that *Drosophila* eye color is linked to alleles present on the sex chromosomes.

Morgan also noted that sometimes not all the alleles in a linkage group were inherited together. At first he hypothesized that mutations had occurred, but mutations are rare, usually occurring in only one individual out of tens of thousands. Consequently, he inferred that the alleles had somehow rearranged themselves through *crossing-over*.

301

Bio Buzz

Crossing-over is the exchange of alleles between homologous chromosomes.

Crossing-over occurs more frequently between some alleles than others. All alleles occupy a fixed position on the chromosome, and those that are closer together on the chromosome are less likely to be separated by crossing-over. One of Morgan's students, Alfred H. Sturtevant, collected enough crossing-over data to construct a chromosome map, a device that diagrams the position of each allele position on a particular chromosome.

Although Morgan had started with classical genetics, much of what he discovered and ended up studying could not have been predicted by Mendelian inheritance principles. What Morgan was engaged in would foster a whole new branch of genetics, called molecular genetics.

Population Genetics

It's interesting to track one or two pairs of alleles over several generations: The predictability ratios tend to remain stable from generation to generation. But even if you're tracking only a single gene through the generations, the resulting phenotypes may not always be what you expect.

Weird Science

Environmental conditions can influence the expression of genes. In one experiment, researchers plucked a patch of hair on the back of a Himalayan rabbit (which is normally white, with black paws, ears, and nose) clean of hair. (The rabbit's back contained nothing but white fur.) They then covered the bare skin with an ice pack. When the hair grew back, the hair was black, not white. Without an ice pack, or lowered temperatures, the hair grew back white.

Variations of expression exist within each gene. Within a population, individuals will display a range of small differences in the phenotypes of most traits. This population characteristic is referred to as continuous variation and mostly occurs because of the number of genes that affect the trait, as well as environmental influences.

Think about your own eyes. Their color is the cumulative product of generations of gene interaction. Yet, blue eyes are not all the same, nor are brown eyes. Likewise, the height and size of the individuals of any population also vary. What is the source, or the reason for all these variations?

Remember that most organic compounds are synthesized through a series of metabolic actions, and different enzymes regulate different steps. Each of these enzymes is the product of a gene. Scientists speculate that some enzymes can cause a gene to mutate several different ways. Or, perhaps the enzyme product of one gene blocks the actions of another, causing it to run nonstop or to stop its processes too soon. Nutrition or environmental factors also can influence the production of enzymes and other organic processes. Any of these factors could be capable of influencing the expression, or phenotype, or a particular genetic trait.

The field of genetics is wide open for research. Classical, Mendelian genetics, statistical analysis, and population studies have produced a staggering amount of information, but realistically, very little is known about the genotypes of most organisms. Several ongoing research projects may produce more information, but right now, if you're looking for a new career, genetic research is it.

We've covered a lot of abstract stuff in this chapter, although you can definitely see the end results of genetic traits: Just look at yourself in the mirror. We've talked about how genetic traits are transferred from parents to offspring and how some traits are more prevalent or dominant than others. In the next chapter, we'll be discussing genetic engineering, or how scientists are learning how to manipulate and change genes.

The Least You Need to Know

➤ Genetics is the study of how traits are transferred from parents to offspring through inheritance and heredity.

➤ Gregor Mendel was the first scientist to discover and prove that offspring are the products of specific units transferred to them by their parents, not an ambiguous blending of parental traits.

➤ Parents transfer traits to their offspring through genes. A gene is the segment of DNA on a chromosome that controls a particular hereditary trait.

➤ Within any population, there will be individual variations of specific genes. Variations are caused by many factors, such as enzyme interactions and environmental influences.

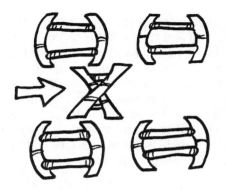

Chapter 24

Genetic Engineering

In This Chapter

➤ Recombinant DNA and biotechnology

➤ Cloning

➤ DNA fingerprinting

➤ The Human Genome Project

As you discovered in the last chapter, Gregor Mendel's experiments were a deliberate manipulation of genetic material, albeit on a pretty tame scale. By mixing and manipulating the genes of pea plants, Mendel was able to produce new plants and formulate his own genetic principles.

Today, genetic manipulation has gone way past Punnett squares and pea plants. While providing solutions to many medical problems and advancing agriculture, the new fields of biotechnology and genetic engineering also raise ethical, social, legal, and ecological questions. Will the benefits outweigh the risks? That's the subject we'll discuss in this chapter.

Recombinant DNA and Biotechnology

Nature has been conducting her own genetic experiments for at least three billion years. Through crossing-over, mutations, and any other number of events that alter the transmission of genes, nature has been producing all sorts of anomalies on her own.

The fossil record and evolutionary timetable bears witness to this. Every species on the planet has changed and developed through an age-old evolutionary process triggered by the alteration of genes by some internal or external force.

We humans have been performing our own genetic alterations to a variety of other species for centuries, certainly long before the availability of current technology. Through artificial selection processes, we've produced new types of cattle, new crops, and any number of "designer" dogs and cats. We've developed larger, sweeter oranges, and even meatier poultry with more white meat than dark. We've even managed to grow watermelons without seeds and create some unusual hybrids of species, such as the mule (horse × donkey) and the tangelo (grapefruit × tangerine).

While Mendel's methods of genetic analysis are still widely used, especially in the field of Mendelian, or classical, genetics, nowadays, researchers use *recombinant DNA technology* to analyze most genetic changes.

A scientist working with recombinant DNA technology cuts and splices together genes from different species, and then inserts these modified molecules into a type of cell that engages in rapid reproduction, usually a bacterial cell, such the *E. coli*. Because bacteria typically engage in rapid reproduction and cell division, the cells copy the foreign DNA right along with their own. In a very short time, researchers can obtain huge populations of bacteria, each one producing usable quantities of recombinant DNA.

Recombinant DNA technology is the basis for the modern field of *genetic engineering*. In genetic engineering, genes are isolated, modified, and then reinserted back into the original organism or a different one. By changing a cell's DNA, you change the way it replicates and reproduces. For instance, if a particular human illness is genetically caused, if you can change the DNA in the gene responsible for the illness, you can possibly eliminate the illness.

Plasmids and Restriction Enzymes

Believe it or not, this fancy technology starts with lowly bacteria. Bacteria cells contain only a single chromosome, one circular molecule of DNA. This single chromosome contains all the information the bacterium needs to grow and thrive, but many bacteria

Weird Science

Luther Burbank was an American plant breeder of the early twentieth century who achieved enormous success with selective breeding practices. Burbank grew large amounts of fruits, vegetables, flowers, and grains. Out of his crops, he selected only those with traits he considered desirable, and then bred them further. He ended up developing more than 800 new plant varieties, including the spineless cactus and the seedless grape.

Bio Buzz

Recombinant DNA technology involves procedures in which individual genes within a strand of DNA from different species are isolated, cut, and spliced together; the new recombinant molecules then are reproduced in quantity.

species also contain superfluous molecules of DNA, called *plasmids*. Each of these DNA molecules of DNA contain a few genes.

The genes in plasmids, and plasmids themselves, aren't essential to the bacterium's survival, but some contain genes that transfer beneficial traits to the bacterium. For example, the genes in some plasmids give the bacterium resistance to antibiotics. The greatest benefit of plasmids for genetic engineering is that their replication enzymes copy and reproduce the plasmid DNA in the same way that they would copy chromosomal DNA.

In nature, many bacteria engage in their own recombinant technology. Many bacteria are able to transfer plasmid genes to neighbors of the same species or even different species. Replication enzymes in the recipient bacterium then integrate the new plasmid into their own bacterial chromosome. The result is a recombinant DNA molecule.

Viruses also engage in gene transfers and recombinations. Just as viral infections don't do humans any good, neither do they offer any benefits to bacteria. Bacteria have evolved an excellent method of keeping out unwanted genes, through the use of restriction enzymes, a special type of cellular protein.

Restriction enzymes are vital to genetic engineering. They recognize and cut apart any foreign DNA that enters a cell. By isolating and controlling restriction enzymes, scientists have their own molecular scissors. When used together, plasmids and restriction enzymes are a major component of genetic recombination under laboratory conditions.

How Restriction Enzymes Work

Each type of restriction enzyme recognizes a very short, very specific nucleotide sequence in DNA. When it recognizes its segment, the restriction enzyme makes a cut, producing a fragment. Because different restriction enzymes recognize different sequences in DNA molecules—and only those sequences—they can be used with precision. Different enzymes can be used to isolate and segregate very specific human genes.

Because DNA is double-stranded, the restriction enzyme needs to cut two identical sequences to produce the fragment. The sequence on each strand might be different, so when the restriction enzyme makes the cut, it's staggered—there's a single-stranded portion at each end of the fragment. These strands are referred to as sticky ends.

The end strands aren't sticky, per se, but they have the chemical capacity to pair up with any other DNA molecule that has been cut by the same restriction enzyme. Suppose that you use the same restriction enzyme to cut both plasmids and isolated DNA molecules from a human cell. If you mix the plasmids and human cell DNA molecules together, they will bind together at the cut sites, which are the "sticky" ends.

The final step is to add *DNA ligase* to the mixture. DNA ligase is an enzyme that produces the same results as DNA replication: It seals DNA's sugar-phosphate backbone at the cut and newly bonded sites.

By following this process, scientists are able to create recombinant plasmids. But these are plasmids with a difference: They now contain pieces of DNA from another organism. With these plasmids, scientists can create a *DNA library*.

Bio Buzz

A **DNA library** is a collection of DNA fragments, all produced by plasmids and restriction enzymes.

Think About It!

Many bacteria, even the kind that causes disease, can be helpful under the right circumstances. The same *E. coli* that causes a nasty, sometimes fatal gastrointestinal disease is frequently used in genetic engineering because it reproduces so rapidly. *E. coli* divides every 20 minutes, so in very short order, you can recreate large amounts of your DNA library.

Because any DNA library is minuscule, scientists must find a way to amplify it, and this is where bacteria really enters the picture. The quickest way to amplify a DNA library is to use "factories" of yeast, bacteria, or any other type of cell that can take up the plasmids and replicate them rapidly.

Once they get enough fragments to work with, scientists sort them by length, using a technique called *gel electrophoresis*. During electrophoresis, an electric field forces the molecules through a viscous gel. Large fragments can't move as fast as smaller ones, so they separate from one another in the gel.

After the fragments have separated, they're put in a buffered solution, and more electrical power is applied. The voltage causes the fragments to migrate toward the charged pole, but again, they do so at different rates because of their different lengths. The fragments further separate into bands, also according to their length. After a specified time of electrophoresis, the staining of the gel can identify fragments of different lengths. With the fragments grouped according to length, researchers then set about determining the exact order of the bases in each fragment of DNA. A complex instrument called a DNA sequencer is used, which determines the nucleotide sequence of each fragment.

By using gene sequencing, scientists have recently been able to discover the genes that are responsible for cystic fibrosis, as well as Duchenne muscular dystrophy. Identifying these genes and the defective proteins that they code should make it easier for scientists to design specific gene therapies to correct the actual defect.

Double Trouble!

Lots of images come to mind at the word *clone*, most of them conjuring up visions of mad scientists cackling over test tubes. Biologically speaking, though, there are almost as many definitions to go with the word as there are images.

For starters, there are at least three pertinent biological definitions for the word *clone*.

1. Clone: A group of genetically identical individuals, all descended from the same parent through asexual reproduction. (Remember coral polyps and plants that produce suckers? They're all clones.)

2. Clone: A group of genetically identical individuals, all produced by mitotic division from an original cell. In this type of cloning, the cell creates a new set of chromosomes and then splits into two daughter cells. (Your own body engages in this type of cloning all the time; it's how your body replaces worn-out cells.)

3. Clone: A group of DNA molecules produced from an original length of DNA sequences, produced by bacteria or viruses, using molecular biology techniques. This is the high-tech stuff, and it's usually referred to as molecular cloning, or DNA cloning, to distinguish it from other types of cloning.

Cloning is a natural biological process; it happens all the time in nature. And when it comes to humans, there are already thousands of clones on Earth. Technically speaking, identical twins are clones because they start out as a single egg, which then divides in two. Identical twins share the same DNA. But molecular cloning is something new. It doesn't occur in nature, and the implications of its development and use are mind-boggling. In recent years, this type of cloning has zoomed to the forefront of biological research, as well as newspaper headlines.

In 1997, scientists successfully cloned the first mammal. Dolly, an ordinary-looking lamb from Edinburgh, Scotland, found herself splashed across the headlines. She may have looked ordinary, but she was unique because she was an identical copy of an adult sheep, and she had no father. Even the sheep that gave birth to Dolly was a surrogate mother; scientists implanted the embryo that became Dolly in her uterus.

Think About It!

What do you think about molecular cloning? Do you think we should explore all the possibilities of this newly emerging science? Or what about the idea of humans being able to clone themselves or to "design" new lives by manipulating DNA? If you haven't thought about it yet, you'd better start thinking now. In many countries, including the United Kingdom, the cloning of humans is illegal, but that's not the case in the United States.

Weird Science

For all the furor surrounding Dolly, molecular cloning is not just right around the corner. For one thing, nuclear transfer is labor-intensive and very costly. For another, it frequently fails. Dolly was the result of 277 nuclear transfers—it took that many to produce a single, viable cloned lamb. At present, cloning from embryonic cells—and even old-fashioned animal breeding—is a more efficient way to produce large amounts of genetically altered animals.

Other animals had been cloned before, but they'd been cloned from embryonic cells. Embryonic cells are totipotent, which means that each cell has the potential to become an embryo. But while the embryo is still in the womb, some genetic switches get flipped and the cells lose their ability to generate an embryo, they become any number of specialized cells: skin, eyes, liver, and so on. This is convenient—when you cut yourself, it's handy to have your skin produce scar tissue, and not, say, a toenail.

What makes Dolly so unique is that scientists didn't clone her from embryo cells (which isn't an easy process itself), but from the udder cells of a 6-year-old ewe. Researchers at the Roslin Institute in Scotland found a way to make fully differentiated adult cells become totipotent again.

Dolly also wasn't cloned through standard cloning procedures. She was cloned through the technique of nuclear transfer. In a nuclear transfer, scientists remove the DNA from an unfertilized egg cell and substitute it with the nucleus of a specially treated differentiated body cell. An electrical pulse or chemical compound is then used to fuse the unfertilized egg "shell" and the new nucleus into a single new cell, which, hopefully, will develop.

As far as Dolly was concerned, the biggest breakthrough of the Roslin Institute was that it had found a way to make a differentiated adult cell become totipotent again. They did so by starving the isolated cell, which makes it shut down. It stops producing proteins from its own DNA and then stops trying to grow. This state is referred to as G0. When the cell to be cloned reaches G0, researchers implant the nucleus, which contains its DNA, into the denucleated egg.

Dolly is currently the world's most famous sheep, but in the field of cloning, many nameless, faceless animals who are clones are also noteworthy. The main thrust so far in cloning animals has been to produce tissues that can benefit humans. For instance, Polly, another sheep cloned at the Roslin Institute, has had a human gene spliced into her DNA. Because of this, she produces milk that contains Factor IX, a blood-clotting agent needed by men with hemophilia B.

Aside from the ethics of human cloning, some other considerations are involved, including aging. Just as your body ages, so does your genetic DNA. By the time you're an adult, your mitochondrial DNA has been damaged and repaired many times. One of the current theories about aging is that when your mitochondria run down, your cells die. This means that by using "old" cells from adult animals, scientists might be creating "old" animals, even if they look like babies. They may very well suffer an early and accelerated aging, or inexplicably die when very young. So far, though, Dolly seems fine.

Genetic engineering is the fastest-growing field of science. Indeed, the coming millennium has been referred to as the Age of Biology. Advances in molecular engineering and cloning are fast outstripping our ability to deal with their consequences. Legal or not, human cloning does not require a lot of equipment. The techniques pioneered at the Roslin Institute demand great skill and patience, but the required equipment is commonplace in most laboratories. Researchers from the Oregon Regional Primate Center in Beaverton, Oregon, already have announced that they have cloned monkeys from embryonic cells. And researchers have long known how to apply similar techniques to clone cows and rabbits. So far, no human has been cloned, but it's probably just a matter of time.

Got You By the Riff-Lips: DNA Fingerprinting

Nothing says *you* like your DNA. Humans and other sexually reproducing species have a unique array of *RFLPs* (pronounced "riff-lips") that they inherit from each parent in a Mendelian pattern. They're yours and yours alone.

Let's assume that a researcher has a good-sized sample of your DNA (obtained from any part of your body—blood, semen, hair, saliva, skin cells, or even from a piece of your toenail someone picked up off the floor).

The researcher then cuts your DNA with restriction enzymes, subjects the fragments to electrophoresis, and sequences them. When the researcher has gotten this far with your DNA, he'll apply the Southern blot method. The Southern blot method is a technique by which the electrophoretically separated fragments are transferred to a membrane, which is then immersed in a solution containing a *labeled DNA probe.*

Weird Science

Transgenesis, the process of transferring genes from one species to another, is not new. Transgenic animals, especially mice, are often used as models for disease study. Most transgenic animals are created through some form of a knock-out. A knock-out occurs when a particular gene is inactivated, often by inserting another gene that disrupts the operation of the target gene. Transgenic animals are being developed as bioreactors—they are bred to produce pharmaceutical products in their milk. Besides treating hemophilia, transgenic animals are also being bred to treat cystic fibrosis.

Bio Buzz

RFLP stands for restriction fragment length polymorphisms. RFLPs are the differences between DNA electrophoresis patterns.

311

The labeled DNA probe will bind to whatever portion of your DNA the researcher is interested in. Now suppose that the researcher applies all these techniques to the DNA from *other* people. Pretty soon, there will be several DNA samples in front of the researcher, all with different variations in the banding patterns. Why? Although the DNA of all humans has many commonalities, the molecules of any two people still differ significantly. These molecular differences cause the number and location of sites where restriction enzymes can cut to also differ. And they differ just enough to give each person a unique DNA fingerprint.

RFLP analysis is an amazing procedure with some wide-ranging and startling implications. The most obvious one, of course, is with crime detection. Britain started the first DNA database in the world and thus is now heads above everyone in solving crimes through DNA fingerprinting. RFLP analysis was the primary tool used to identify the remains of Russia's imperial Romanoff family, who were secretly shot in 1918. The remains of five bodies found in a shallow grave were positively identified through molecular genetic techniques. This same analysis also confirmed that the bodies of two of the children, Alexis and Anastasia, were not among the group.

Bio Buzz

A **labeled DNA probe** is any short DNA sequence that scientists have synthesized from radioactively labeled nucleotides. Part of the probe must be designed to pair with the portion of interest in the DNA that's being examined.

Try It Yourself

Over the centuries, science has advanced because people thought and pondered things of interest. Many questions are not immediately answerable, and some may never be. Put on your thinking cap and see if you can advance the scientific cause.

In addition to any examples we've listed in this chapter, what are some other potential benefits of genetic engineering? What are some other potential risks of genetic engineering?

Nowadays it's not uncommon to find genetically engineered food in your supermarket. Most of these food items (tomatoes, for instance) look better than "regular" food items. Given the choice between buying genetically engineered food, or not, which would you buy? Why?

In Britain, police can lawfully collect DNA samples from suspects—by force, if necessary. In one case in Cardiff, Wales, more than 5,000 "donors" gave DNA samples in an effort to bring the rapist and murderer of a local teenager to justice. Mass DNA screening has already solved several murders in Britain, Germany, and other European countries.

The United States lags far behind other countries in using DNA fingerprinting for criminal prosecution. A few years ago, in a high-profile criminal case, a celebrity was charged with murdering his ex-wife and one of her friends. The murderer had left blood at the crime scene, as well as a trail of blood that led to the celebrity's house. RFLP evidence strongly implicated the celebrity. However, his lawyers hammered away at the competency of everyone who had ever been at the crime scene or had dealt with any part of the testing, and the RFLP evidence was discounted.

RFLP analysis also has long-range potential for medicine. Unique restriction sites already have been pinpointed on several genes, including the ones that cause cystic fibrosis, sickle-cell anemia, and other genetic disorder diseases. In finding the sites, researchers are better able to formulate prenatal diagnoses and gene therapy plans.

Evolutionary biologists are also using RFLP analysis to decipher the genetic codes of mummies, fossilized plants, and insects. Humans that have been preserved in peat bogs for thousands of years, and wooly mammoths buried in ice are also being investigated.

The Human Genome Project

The Human Genome Project is an ambitious undertaking started by the U.S. Department of Energy and the National Institutes of Health in 1990. The Human Genome Project has five main goals:

1. To identify all of the nearly 100,000 genes in human DNA
2. To determine the sequences of the nearly three billion chemical bases that make up human DNA
3. To store all this information in databases available to everyone
4. To develop tools for genetic data analysis
5. To address all ethical, legal, and social issues (referred to as ELSI) that arise from the project

Weird Science

The Human Genome Project is the first large scientific project that is attempting to address ethical, legal, and social issues arising from the work. About 3 to 5 percent of the annual budget for the project has been set aside to study these issues exclusively.

The Human Genome Project also includes parallel studies of bacteria, such as *E. coli*, laboratory mice, and fruit flies. All organisms are related, in that they share similar DNA sequences. By identifying the genetic maps of these "simpler" organisms, scientists will have further understanding of the human *genome*.

If you recall, DNA and genes carry all the information an organism needs to make proteins. The protein composition of an organism determines many things, including how the body metabolizes food and fights infection, and even how the organism looks and behaves.

Bio Buzz

A **genome** is the complete set of instructions for making an organism. The human genome contains all the DNA in a human being, including its genes.

Weird Science

The human genome is so enormous that to catalogue it in books would require 200 volumes the size of the Manhattan phone book (which is 1,000 pages). If you read the three billion bases out loud without stopping, it would take you 9.5 years to finish. And this is calculated at the reading rate of 10 bases per second, to equal 600 bases per minute, 36,000 bases per hour, 864,000 bases per day, and 315,360,000 bases per year.

DNA is composed of four similar chemicals that are repeated millions and billions of times throughout the genome. These chemicals are called bases, and the four are referred to by abbreviated form: A, T, C, and G. The human genome has three billion pairs of these bases.

The order of the bases is very important because it determines the diversity of life. The order of your bases determines that you are human, as opposed to rice. Everything—fruit flies, giraffes, wheat, grasshoppers—has its own unique genome and base ordering.

Obviously, the scope of the Human Genome Project is enormous. The medical advances are expected to be staggering, and the whole project itself is spurring technological advances. Just storing all the information requires technology previously unheard of.

Although the United States is heading the project, many countries are taking part. At least 17 different countries have established major genetic research centers to engage in the Human Genome Project. These larger programs are in Australia, Brazil, China, Denmark, the European Union, France, Germany, Israel, Italy, Japan, Korea, Mexico, Netherlands, Russia, Sweden, and the United Kingdom. Several less-developed countries are participating by conducting studies of molecular biology techniques, and by studying the genomes of organisms that are pertinent to their geographical locations.

The original timetable for the Human Genome Project was to have all human genetic material identified and mapped by the year 2005. However, because of rapid technological advances, scientists involved expect to complete the project as early as 2003.

In December 1999, researchers at the Sanger Centre in England; the University of Oklahoma; Washington

University in St. Louis, Missouri; and the Keio University in Japan announced that they had succeeded in deciphering the sequence of 33.5 million "letters," or chemical components, that make up the DNA of chromosome 22. This was a true milestone because it was the first time in history that any scientists had ever unraveled the genetic code of an entire human chromosome. As of March 2000, 64.1 percent of the three billion DNA base pairs had already been identified.

After deciphering the human genome, the next task will be to figure out what it all means. For instance, chromosome 22 alone has already been implicated in the workings of the immune system, congenital heart disease, schizophrenia, mental retardation, several birth defects, and several cancers, including leukemia.

As the Age of Biology commences, humans will be dealing with more rapid technological advances than they have ever seen. Ten years from now, how will DNA fingerprinting figure into justice systems? How many pharmaceutical products will transgenic animals be producing? And what will we have done with our complete knowledge of the human genome?

We've only touched the surface of genetic engineering, but we've still covered a lot of ground. We've talked about the unique patterns that exist in each person's DNA, and we've discussed how DNA can be re-engineered and then reinserted into the cell. We've also explored the Human Genome Project, an ambitious international undertaking that is attempting to map every gene in the human body. In the next chapter, we'll discuss the field of science that ties all living organisms together: ecology.

The Least You Need to Know

➤ Recombinant DNA technology involves procedures in which genes in DNA from different species are isolated, cut, and spliced together, and in which the new recombinant molecules are then reproduced in quantity.

➤ Recombinant technology is the basis for the field of genetic engineering. In genetic engineering, genes are isolated, modified, and then reinserted into the original organism or a different one.

➤ The DNA of humans contains many similarities, yet each human has a unique DNA fingerprint. The variations that make DNA fingerprints unique are caused by molecular differences in each individual.

➤ The Human Genome Project is an international effort, headed by the United States, to identify all 100,000 genes in human DNA as well as the sequences of the three billion chemical bases found in human DNA. The project also will address ethical, legal, and social issues and will develop new tools for data storage and analysis.

Chapter 25

Tying It All Together: Ecology

In This Chapter

➤ At home on Earth

➤ Living on land

➤ The watery domain

➤ Putting the pieces together

➤ Ever-changing populations

Nearly every part of Earth supports life. From the highest mountain to the deepest part of the sea, life is everywhere. Much of the diversity we see among organisms is the result of the wide variety of climate and geography found on Earth.

Every place on Earth, organisms exist together. No organism exists alone. Every single organism we've discussed—bacteria, humans, eagles, corals, rhinos—is part of an intricately balanced system of both living and nonliving elements.

In this chapter, we'll be talking about the different ecosystems found on Earth, as well as how they all work together.

Living on Earth

Ecology is the study of the relationship between organisms and their environment. The field of ecology is wide open to virtually anything you could want to study because scientists study all aspects of a relationship: how the organisms react together, how groups of organisms react with other groups, how the environment spurs adaptations—the list is nearly endless.

Weird Science

The biosphere of our planet is a tiny, tiny segment of the planet as a whole. If the Earth were the size of a basketball, the biosphere would be thinner than a sheet of paper.

Weird Science

For decades, Isle Royale, an island national park in Lake Superior, has been a living laboratory for the study of all sorts of ecological issues, particularly those of populations and communities. The moose constitute one population, and the wolves that prey upon them constitute another. Together, they make up a community that is perfect for studying interactions. Because Isle Royale is an island, outside influences are kept to a minimum, and both populations are "trapped" together in a roughly 9-by-50-mile area. That's a scientist's dream.

The biosphere of our planet extends into the atmosphere approximately 8 kilometers, and down approximately another 8 kilometers to the bottom of the ocean. All life as we know it exists within this domain. The spores of a few microorganisms and plants, as well as some insects, have been found 8 kilometers above the Earth. And brittle stars and sea urchins can exist a bit deeper than 8 kilometers in the ocean. But most life on Earth exists either on or within a few meters of the Earth's surface.

The biosphere may be thin, but it's hardly small. It's the size of the entire planetary surface and then some because it encompasses the depths of the ocean and a good portion of our atmosphere. To be able to study something this large, scientists study smaller and simpler units within the biosphere.

Ecologists study more than just organisms and how they interact with each other. They also study ecosystems, ecological units that are associations of organisms and their physical environment, all of which are interconnected through a continuous flow of energy and material cycling. Ecosystems start with individuals. Individuals of the same species make up populations. Populations of many species make up a community. The community, together with the nonliving components such as the rocks, soil, and climate, make up the ecosystem.

A lake (and all the organisms inhabiting it) is an ecosystem all by itself, as is a particular prairie or meadow in a mountainous basin. Ecologists break the components of ecosystems into two main categories. All the nonliving components of an ecosystem are called abiotic factors, and all the living components of an ecosystem are referred to as biotic factors.

Abiotic factors include such things as sunlight, temperature, precipitation, chemistry of the atmosphere, chemistry of the soil, and the slope and drainage of the land. These things are not alive (at least, not by our standard definition), but they definitely interact with each other to form the ecosystem. For instance, heavy rains can cause a massive runoff of topsoil into a river. The river will overflow its banks and deposit a layer of silt and sediment over everything it floods. When the river recedes, the landscape will have changed. Even earthquakes and volcanoes are abiotic factors of an ecosystem.

Biotic factors include such things as plants, animals, fungi, protists, and anything else that is alive. Biotic factors also interact with each other. Within any given population, individuals compete for food, water, space, and any other resources in the ecosystem. Within communities, populations compete with other populations for the same resources. For instance, two different species inhabiting the same ecosystem may eat the same plants.

Try It Yourself

You can get a firsthand view of how abiotic factors affect an ecosystem if you visit Yellowstone National Park. The fires that raged throughout the summer of 1988 drastically altered the entire ecosystem. More than a dozen years later, the landscape is nowhere near what it used to be.

Two other biotic factors are predation and symbiosis. In predation, animals of one species (the predators) kill and eat animals of another species (the prey). Symbiosis is a close and usually mutually beneficial relationship between two different species, virtually the opposite of predation.

The abiotic and biotic factors of an ecosystem interact with each other. For instance, rain, sunlight, and soil conditions affect plant growth. This, in turn, affects the distribution of animals, as well as their adaptations. The deer that live in the deciduous forests of the northern United States have teeth and digestive systems that are adapted to eating the leaves of moisture-gathering trees. These same deer would not do well on the grass plains of Siberia, where the plant and animal life is adapted to a harsh, dry, cold climate.

Terrestrial Biomes

The biosphere is enormously varied. The ocean is a continuous body of water that covers 71 percent of our planet. Its temperature and currents influence what happens in the atmosphere. Continents and islands break up the ocean and also influence what happens in the atmosphere.

All these variations in the biosphere—temperature fluctuations, currents, land masses, precipitation, and so on—create their own variations in different parts of the globe. A terrestrial *biome* is a large region of land that can be characterized by the vegetation of the ecosystems found within it. There are seven major types of terrestrial biomes: polar, tundra, coniferous forest, deciduous forest, grassland, desert, and tropical rain forest, as well as a few other more specific categories.

Polar biomes are found at the Earth's coldest extremes: the poles. Polar biome conditions also exist on the top of the world's highest mountains. Polar biomes are characterized by ice. They receive almost no precipitation, and fresh water is scarce. Conditions in these biomes are harsh; the wind relentlessly scours the landscape, and sunlight is nonexistent during the winter months. Yet life not only exists in polar biomes, but it flourishes at every opportunity. In the Arctic polar biome, more than 100 species of flowering plants, as well as mosses and lichens, are present. Mammals, such as polar bears, walruses, seals, and numerous bird species also make their home in the Arctic polar biome for at least a portion of the year.

Very little inhabits the interior of the Antarctic polar biome; only bacteria and some insects. In this biome, most life exists along the coast. Only three species of flowering plants can be found on the Antarctic coast, but seals, whales, penguins and other birds also inhabit the area.

The tundra biome forms a continuous band across northern Europe, Asia, and North America. Tundralike conditions also can be found on the highest mountains, just above the tree line (the highest elevation at which trees can grow). The tundra biome is treeless year-round. During the winter months, it is blanketed with snow. During the summer season (a mere eight weeks), it becomes a patchwork of bogs, ponds, and soggy soil.

Although the summer season is brief in the tundra biome, the area explodes with life. Hundreds of flowering plants burst into bloom. Shrubby plants appear, accompanied by swarms of biting mosquitoes and black flies. Birds appear by the thousands, nesting and bearing young in the brief time span. Large mammals, such as caribou, wolves, and grizzlies, also inhabit the tundra biome.

Bio Buzz

A **biome** is a geographic area that can be characterized by the specific animals and plants that live there, as well as the climate conditions that exist. Within a biome, there may be any number of ecosystems.

Think About It!

In addition to having no trees, the tundra biome is characterized by the presence of permafrost, a permanently frozen layer of ground more than 500 meters thick. Even when the snow melts, the water can't penetrate the permafrost, hence the soggy summer conditions.

On the planet, the coniferous forest biome lies just south of the tundra biome. Coniferous forest biomes are dominated by (surprise!) conifers, cone-bearing trees such as pines, firs, cedars, and spruces. The plants that live in this biome are adapted to relatively harsh conditions: long winters, short summers, nutrient-poor soil, and frequent dry periods. The leaves of conifers are covered with a waxy cuticle and remain on the trees all year. Large mammals, such as moose and bears, as well as smaller critters reside in coniferous forest biomes. Many of these animals rely on stored body fat to get them through the winter months. On the other hand, most invertebrates species hibernate during the winter months.

The deciduous forest biome is found in more varied regions of the globe. Deciduous forest biomes are characterized by the presence of trees that lose their leaves in the fall. These types of biomes are found throughout much of Europe and the United States, as well as parts of South American, Australia, Asia, and Africa. These regions of the globe tend to have pronounced seasons (usually four—spring, summer, fall, and winter), with precipitation distributed evenly over the year. These biomes are also characterized by nutrient-rich soil and a stunning diversity of plant and animal species.

Deciduous trees usually have a large surface that permits maximum light absorption. Birches, beeches, oaks, maples, cottonwoods, willows, and hickories are all deciduous trees. Deer, foxes, raccoons, and squirrels typically inhabit deciduous forests.

Deciduous forests are home to hundreds of bird species, although the majority of these species are just passing through. Most of them arrive to spend only the summer and then migrate south for the winter months. About one-fourth of all the bird species that live in deciduous forests, though, stay there year-round. Cardinals and blue jays are just two such species that can be found in deciduous forests during the winter months.

The grassland biome is dominated by numerous species of grasses. Grassland biomes occur at approximately the same latitude as deciduous forests, but they experience more variation in temperature ranges and don't receive enough rainfall to support trees.

There is no common name for the various grasslands that dot the planet. The grasslands of the United States are called the prairie. Then there are the steppes of Asia, the pampas of South America, and the veldt of South Africa. The savannah of Africa is a grassland, but unlike other grasslands, it's also scattered with subtropical and tropical trees.

Grassland biomes produce a lot of food and support a lot of animals, especially grazing animals such as bison and antelope. In South America, the rhea, a large flightless bird, is a major grazer.

Weird Science

Grassland biomes cover nearly one-fourth of the land surface of Earth, making them the largest biome type.

Elephants, antelope, and giraffes in Africa, and kangaroos in Australia inhabit grassland biomes. Billions of insects also inhabit the grasslands, happily munching away.

Desert biomes are found around the world and are characterized by a lack a rain—less than 25 centimeters a year. Yet most deserts harbor numerous species of plants and animals. When you hear the word *desert*, you tend to think of bone-parching heat, but not all deserts are hot year-round. For instance, during the winter months, temperatures in the deserts of Nevada and western Idaho frequently plunge below freezing. In the Sahara Desert in Africa, rain may fall only once every few years.

Like all other organisms on Earth, desert plants and animals are well adapted to their environment. Cacti have evolved specialized root and water storage systems, and have learned to reproduce rapidly if any precipitation occurs. Within days—and sometimes hours—of rainfall, plants in any desert biome will burst into bloom, with each plant taking advantage of the unexpected boon of water. Not many large mammals inhabit desert biomes, but numerous reptiles, birds, insects, and smaller mammals thrive. Many are nocturnal, staying hidden in cool underground burrows during the heat of the day.

Think About It!

For an interesting perspective on life in the desert, read Edward Abbey's classic book *Desert Solitaire* (Ballantine Books, 1968). This is Edward Abbey at his best, speaking about the landscape that he loved.

Try It Yourself

Visit your very own rain forest. Although rain forests sound exotic to most Americans, we have one right here on our own continent: the rain forest biome, called the temperate rain forest. This particular biome extends from central California up to southern Alaska and is characterized by large, diverse plant life, such as the giant redwoods and the Sitka spruce. The mammals of the temperate rain forest, creatures such as deer, elk, and rodents, live on the ground, unlike their tropical rain forest counterparts.

The tropical rain forest biome is found near the Earth's equator because this global region is where the greatest amount of sunlight and rainfall is found. Tropical rain forests can get as much rain in one month as some grasslands get in a year.

Tropical rain forests contain some of the most diverse plant and animal life found on Earth. One small area may contain more than 100 different species of plants, some as high as 45 meters, while others creep along the forest floor. Although most people think of tropical rain forests as impenetrable jungles, the forest floor is usually relatively free of vegetation because not much sunlight reaches it.

Animal life is also incredibly diverse. Hundreds of thousands of insect species live in tropical rain forest biomes. Birds, primates, reptiles, and large and small mammals are also abundant. Many tropical rain forest animal species spend most of their time in the trees, essentially living there rather than on the ground.

Aquatic Biomes

The majority of all biomes are aquatic. Aquatic biomes encompass everything from the Pacific Ocean to the Nile River, to a half-acre woodland pond.

The marine biome is the planet's largest biome. (This is hardly surprising, given that 71 percent of the planet's surface is covered by ocean.) The marine biome is so extensive that, for study purposes, scientists have broken it into three different areas: ocean, intertidal zones, and estuaries. Of these three, the ocean category is still so large that it's subdivided even further.

The ocean itself is broken into two categories: the open ocean, or pelagic zone, and the ocean bottom, or benthic zone. The ocean is also categorized according to the amount of light it receives. The photic zone extends to a depth of about 200 meters. The photic zone is where most photosynthesis occurs. The aphotic zone extends from the bottom of the photic zone down to the ocean bottom. Because little light penetrates to these depths, little photosynthesis occurs.

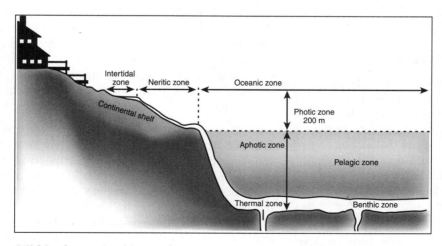

Within the marine biome, the ocean is further categorized into several more zones.

The pelagic zone is still so large that scientists further categorize it by depth and relation to land masses. The *neritic zone* extends from shorelines over the continental shelf. This zone gets the greatest amount of sunlight, and upwelling currents from the deeper regions bring an abundance of minerals. Consequently, the neritic zone supports the greatest amount of marine life. Plankton, all species of fishes, sea turtles, squid, dolphins, and numerous other animals live in or visit the neritic zone.

The second subarea of the pelagic zone is the oceanic zone. This is the deep water of the ocean. Even in the photic zone, the oceanic zone is much less populated than the neritic zone.

The mineral levels of the oceanic zone are too low to support much life, and what life there is looks downright weird (which is saying a lot, given everything we've already looked at). Deep-sea fish have lower metabolic rates and reduced skeletal systems, both of which are adaptations to freezing cold and enormous pressure. Most of these fish have enormous jaws and teeth, and expandable stomachs, which enable them to catch and digest the rare prey they might happen upon.

After the ocean, the intertidal zone is the next largest category in the marine biome. The intertidal zone is created by the incoming and outgoing tides, and organisms that live in this zone must be adapted to both water and air conditions. When the tide is in, clams, mussels, oysters, barnacles, and other organisms filter plankton from the water. When the tide goes out, they retreat into their shells. Crabs and other creatures burrow into the mud to stay moist. Organisms in the intertidal zone also must be adapted to the force of pounding waves because the shoreline (which is essentially the intertidal zone) receives the brunt of the ocean's force.

An estuary is classed as a marine biome but is really a mix of both marine and freshwater. Estuaries are found all over the world, wherever freshwater rivers flow into the ocean. Some examples of estuaries are salt marshes, mud flats, and bays. Estuary communities are full of life. The shallow waters receive plenty of sunlight, and the rivers deposit lots of minerals and nutrients. The drawback is that the constant mix of freshwater and saltwater creates temperature

Bio Buzz

The **neritic zone** is the subarea of the ocean that extends from a shoreline over the continental shelf. The neritic zone receives the greatest amount of sunlight, and the majority of marine life lives in the neritic zone.

Weird Science

Marine animals that live below 2,000 meters are referred to as abyssal organisms. Besides fish, abyssal organisms also include the unique organisms that inhabit thermal vents. Unusual clams, crabs, and worms inhabit these vents, feeding on bacteria that thrive in the 700° Celsius hot springs.

fluctuations and brackish conditions. Like the intertidal zone, estuaries spend part of their day under water and part exposed to air.

Estuaries are vitally important to the health of the marine biome. Salt marshes are often referred to as the nurseries of the ocean because many species of marine animals spend their infancy there. The youngsters thrive in the relatively calm conditions, and, most importantly, adults aren't hanging around trying to eat them. Migrating salmon also use estuaries as a changing station, getting their bodies geared up to go from saltwater to freshwater, or vice versa.

Containing virtually no dissolved salts is what characterizes the freshwater biome. The ocean is about 3.5 percent salt, whereas freshwater usually has less than 0.005 percent salt. The freshwater biome includes such diverse bodies of water as Lake Superior, the Amazon River, mountain streams in the Himalayas or Rocky Mountains, and the half-acre pond in a cow pasture.

Within the freshwater biome, lakes and ponds are broken into two categories: *eutrophic* and *oligotrophic*.

Eutrophic lakes generally have mucky bottoms and murky waters, which indicates the presence of organic matter. Oligotrophic lakes tend to have clear water and sandy or rocky bottoms. Fish inhabit both types of lakes, as well as birds and small mammals such as muskrats and otters.

A river is any body of water that flows down a gradient toward a mouth. Given that definition, a river can be a few kilometers long, or it can be the Nile in Egypt. The degree of slope, or the gradient, is a key abiotic factor in this type of freshwater biome because the slope affects how fast the water flows and, hence, what type of organisms can live there. Slow-moving waters are also richer in nutrients than fast-moving ones.

Think About It!

The Chesapeake Bay, on the eastern shore of North America, is one of the largest estuary salt marsh systems in the world. For a great read about this region, its wildlife (specifically, crabs), and the people that live there, read William Warner's Pulitzer Prize-winning book *Beautiful Swimmers* (Penguin, 1976).

Bio Buzz

A **eutrophic lake** is rich in organic matter. An **oligotrophic lake** contains very little organic matter.

Ecosystem Structure

All these different biomes contain unique ecosystems—billions of them (which is why, if you'd like to be an ecologist, you'll never lack for work). In a single lake, the biotic and abiotic factors come together, creating one unique ecosystem. Likewise, a single lake is an ecosystem unto itself. Each ecosystem is usually its own closed energy loop, in that the majority of its energy is both produced and consumed within the ecosystem.

The three main aspects of ecology that define an organism's relationship to its ecosystem are its habitat, its niche, and its trophic level.

Think About It!

Rivers have their own marsh systems. For an interesting account of one day in the life of a freshwater marsh, read Sally Carrighar's book *One Day at Teton Marsh* (University of Nebraska Press, 1947).

Think About It!

Large forces as in the movement of the Earth, also can affect an ecosystem. Phenology is the study of natural phenomena that recur periodically. For an interesting (and easy) read that follows a year in the life of the Grand Teton–Yellowstone ecosystem, pick up a copy of Frank Craighead Jr.'s book *For Everything There Is a Season* (Falcon Press, 1994).

An organism's habitat is the physical area in which it lives. Because plants don't move, their habitat is typically one place. But the habitat of bald eagles, for instance, ranges from prairies to deciduous forests to coniferous forests. A mouse's habitat might include a field, a garden, and the walls of a house.

Any drastic change in either biotic or abiotic factors will affect a habitat and every organism living in it. A storm can severely disrupt an ecosystem, as can the unexpected arrival of a predator.

A *niche* is a specific way of life for a species. An organism's niche includes such things as its habitat, its feeding habits, its reproductive behavior, and all aspects of its interaction with its own ecosystem.

Niches are what allow many organisms to occupy the same ecosystem. For instance, if five different species of birds live in the same small ecosystem, you might wonder how they could survive—wouldn't the competition for food and space be too fierce to support all species? Not if they all have different niches. One bird might live only high in the trees, feeding on airborne insects. Another might live only in holes within trees, feeding on insects that live in bark. Perhaps another bird lives mainly on the ground.

One way ecologists study the role an organism plays in a specific ecosystem is by studying the energy flow among the different organisms of the ecosystem. The primary source of energy comes from the sun, but from there, it passes from organism to organism. The relationship between what an organism eats and what it is eaten by is referred to as its trophic level. The trophic level of an ecosystem is essentially its feeding level.

Ecologists study trophic levels in terms of producers and consumers. Autotrophs (plants, algae, and so on) are the primary producers of any ecosystem simply because they're the ones manufacturing all the organic nutrients. Heterotrophs are consumers. A primary consumer is a heterotroph that consumes primary producers. A secondary consumer is a heterotroph that consumes primary consumers.

Consumers can be herbivores, carnivores, omnivores, or scavengers. Primary consumers are usually herbivores, while secondary consumers are usually carnivores, omnivores, and scavengers. At the end of all the production and consumption, decomposers recycle the last bit of energy back into an ecosystem. Fungi break down dead tissues and wastes into carbon, phosphorus, nitrogen, and other elements and compounds that then are taken up by the autotrophs to make produce more food—a.k.a. energy.

Population Dynamics

What makes some populations thrive, while others falter or even die off? An increase in population is called population growth. The *population growth rate* is the change in the number of individuals in a specific population over time.

Some populations grow so rapidly that they exhaust the resources necessary for survival. Sometimes populations decline because predators enter the ecosystem. Sometimes abiotic conditions affect the population. Sometimes, all these factors can occur simultaneously, which really wreaks havoc within a population.

Isle Royale, the island national park in Lake Superior, offers a perfect example of a population growing too rapidly for its resources at the same time that new predators were introduced, and at the same time that abiotic conditions affected the population. Around 1910, some moose swam out to Isle Royale (no easy feat—the closest land is about 15 miles away). The island was covered in lush vegetation, and the moose thrived because they had plenty of food and no predators (very few bears and moose can swim 15 miles). They thrived so well that scientists think that at one point there were more than 3,000 moose on the 210-square-mile island: They defoliated the island, and the majority of the population died of starvation.

Gradually, the population stabilized, although the island was no longer so heavily foliated. Then in the winter of 1948, a few wolves—at least two of them a breeding pair—managed to walk across 15 miles of ice during one of the rare years when Lake Superior froze over. Suddenly, an island that had been predator-free had large predators, who proceeded to reproduce.

After initially declining (because they were now being eaten), the moose population again stabilized and the island became more heavily foliated. Then in the early 1980s, a virus started killing off the wolves. In the space of two years, the pack went from 50 to 14. At one point, it dropped to 12. The moose population soared, reaching 2,400 in the summer of 1995. By the spring of 1996, more than half the moose were dead from starvation.

Then the abiotic influences hit. Hard winters and wet, cold springs in 1997 and 1998 caused even more moose to die; the old and the young couldn't survive. In 1999, researchers counted 750 moose and 25 wolves.

For decades, Isle Royale has proved to be one of the best places on the planet for researching population dynamics on all counts. As an island, it also has provided valuable information about ecosystems and how they function. It's such a unique biome that it has been designated an International Biosphere Reserve. That's ecology, all wrapped up in a 9-by-50-mile island.

The field of ecology is large and complex; it's really an entire area of study unto itself. In this chapter, we've touched briefly on the different major ecosystems that are found on Earth, how an ecosystem functions and interacts with other ecosystems, and how populations function within ecosystems. In the next chapter, we'll be talking about the problems besetting our planet and biosphere.

The Least You Need to Know

➤ Ecology is the study of the relationship between organisms and their environment.

➤ The biosphere is the area of our planet where all life exists.

➤ All living parts of an ecosystem—the plants and animals—are its biotic factors. The nonliving parts of an ecosystem are its abiotic factors. A severe disruption of either factor will change the ecosystem.

➤ Both biotic and abiotic factors have a strong influence on whether a population grows or declines.

Chapter 26

The Future of the Biosphere

In This Chapter

➤ Destroying our home

➤ Changing the climate

➤ Overpopulating the planet

➤ Preserving our home

We've been speaking about ecosystems without considering the one single organism that produces the greatest impact on any ecosystem: humans. Humans are capable of disrupting an ecosystem, even entire biomes, more than any other biotic factor.

This disruption has been going on for centuries, if you could truly call it disruption. After all, we live here too; we evolved along with all the other species. But we seem to be the only species on the planet that engages in deliberate destruction of its home. Just as we are the major destroyers, we are also the only species capable of repairing damage to our biosphere.

In this chapter, we'll discuss the ways we are destroying our planet, and we'll explore some possible solutions for reversing this trend.

Resource and Habitat Destruction

All aspects of the biosphere—the atmosphere, the ocean, and the land—interact in ways that control the living conditions on our planet. Starting with sunlight streaming onto the planet, these interactions create worldwide temperatures and circulation patterns that dictate whether life survives or perishes.

Unlike other species on the planet, humans have been altering their environment since they arrived about two million years ago. Humans became key players in the fate of the biosphere, influencing it before they ever knew how the interactions worked or what the consequences would be.

Humans are also prolific: As the human population increases, so do its needs for natural resources such as water, soil, lumber, and fuels for energy. Natural resources are the raw materials that support life on Earth—all life, not just human life. Natural resources can be both biotic and abiotic.

As a species, we live an industrialized life. Our society is based on consumption and growth beyond our physical needs. Over the centuries, our consumption has led to a depletion of natural resources, and subsequent destruction of specific habitats and the biosphere in general.

Bio Buzz

A **renewable resource** is any material that can be regrown or replenished.

Renewable Resources

Many of the natural resources humans use are renewable. Managing these *renewable resources* correctly will take us further into the next millennium.

The depletion and misuse of renewable resources is the greatest folly humans commit against the biosphere because it doesn't have to occur. Take a forest, for instance. An ever-increasing human population has created huge demands for lumber—lumber for homes, paper, fuel, and furniture. Consequently, enormous amounts of trees are being cut down all over the planet in an effort to meet the needs of the human population (cutting down trees clearly doesn't meet the needs of any other species).

Think About It!

Watch your paper consumption, and help save the planet. Deforestation does more than use up a natural resource. At best, deforestation affects entire ecosystems; at worst, it destroys them. Animals are displaced or destroyed. The ground dries up because moisture that was formerly held in the ground by roots systems and shade is now subject to heat and evaporation. With the soil dry and exposed, erosion occurs, destroying aquatic ecosystems and ruining the landscape.

Yet, forests are renewable. For every tree cut down, at least one new one can be planted in its place, thereby replacing the resource.

Unless you're a hunter, you may not give wildlife much thought, other than its novelty value in today's industrialized society. But wildlife is also a renewable resource. Wildlife aside, consider domesticated livestock, including cows, sheep, pigs, and chickens, among others. At one point, all these species were wild in some form or another. Yet humans deliberately turned them into a renewable resource, solely for food consumption.

In the wild, the destruction of any species has an impact on other organisms in the community and ecosystem. Niches disappear, and trophic levels change. Preventing the loss of species is a topic of great scientific concern. The primary threat to wildlife species is not death by human hunters, but loss of habitat. If an organism's habitat has nowhere to live and nowhere to go, it stands a good chance of becoming extinct.

Many countries, including the United States, have laws governing the regulation, protection, and conservation of *endangered species*.

More than 100 countries have signed the C.I.T.E.S. treaty, an acronym for the Convention on the International Trade in Endangered Species. This treaty outlaws the trade of endangered species and their products across international borders. In 1973, the United States passed the Endangered Species Act, which defines endangered species and establishes regulations for their conservation and protection. Most of these treaties and acts do not deal specifically with habitat destruction, though, nor do they really address issues that cause a species that isn't endangered to end up endangered.

Bio Buzz

An **endangered species** is any species that is currently at risk of extinction in its native habitat. Even if there are individuals in zoos, if survival in its native habitat is in jeopardy, the species is still considered endangered.

Weird Science

Habitat destruction ricochets throughout an ecosystem. Destroying the habitat of one organism obviously affects that particular organism. But habitat destruction ripples much further than that, particularly when humans enter the picture. In the 1800s, the Anglian fens in Britain were drained. This changed the habitat from marshland to dry land, which could no longer support the Great Water Dock plant. As the plant became very scarce, the Large Copper, a butterfly species that fed on it, also declined drastically. British butterfly collectors, eager to obtain one of these increasingly rare specimens, captured and killed the few remaining individuals, and the Large Copper became extinct in the 1840s.

It probably doesn't seem like it, but soil is also a renewable resource. It's certainly one that can be conserved with minimal effort or hardship. Soil is continually formed as rocks deteriorate and organic matter piles up, albeit, somewhat slowly. And soil is more than just "dirt"—it's actually a mixture of minerals, rock fragments, and the byproducts and remains of plant and animal life.

331

Humus, the organic component of soil, is the component that enables soil to retain moisture. And humus is constantly being created as part of the natural cycle. A single acre of topsoil, the uppermost layer of soil (composed primarily of humus), can hold up to 11 tons of bacteria, fungi, worms, insects, and other organisms.

All this natural creation does no good, though, if human farming practices keep washing it all away. Every year, more than 2.7 billion tons of soil erode from farmlands in the United States alone. All this eroded soil enters rivers and streams, fouling the water, destroying aquatic wildlife, and clogging stream beds. And that doesn't even consider the agricultural chemicals that are in the eroded soil (which is now in water systems).

To combat erosion, farmers plant grasses or other vegetation on bare fields, put up wind breaks, add plant and animal wastes to depleted soil, and install irrigation ditches, which catch some of the runoff. Contour plowing (following the natural curvature of the land) and terracing also help prevent erosion.

Nonrenewable Resources

Nonrenewable resources require special attention because once they are gone, they're gone—they cannot be replaced. Water, fossil fuels, and minerals are the three main nonrenewable resources that are imperiled by human activity.

Try It Yourself

If you don't think that a dripping faucet wastes water, try this simple experiment. Close the drain in your bathtub, and then adjust the faucet so it drips one drip at a time. Then wait. Before the end of the day, you'll find that your bathtub has filled with water, just one little drip at a time.

Our biosphere contains a fixed amount of water. All the water our planet contains has always been here; no new water is being created. Of all the water on Earth, only 3 percent of it is freshwater available for human consumption. And the small amount of freshwater that's available for human consumption is unevenly distributed. Asia has abundant freshwater and an enormous population that uses that water. Africa, on the other hand, has a large population and very little water. Even in the United States, the water is not always where the greatest population centers are. Southern California is heavily populated, yet it has virtually no water of its own.

Taking quick showers and fixing dripping taps is certainly one way to help conserve water, but it doesn't begin to make a dent in the problem of overall human consumption. For instance, many people consider themselves ecologically minded because they do their part to conserve water and other resources. What they don't take into account is the water that is consumed during the production of goods and services they use and enjoy. As an example, it takes 40 gallons of water just to get *one* egg into the carton you purchase at the grocery store. Clearly, using less water and distributing it more efficiently around the globe is critical to the future of our biosphere.

Huge amounts of water go into the production of many goods and services we take for granted. Take a look at some examples in the chart below.

Consumer Good or Service	Gallons of Water
One egg	40
One pound of flour	75
One loaf of bread	150
One Sunday newspaper	280
One pound of beef	2,500
One automobile	100,000

Minerals are also a finite resource. The minerals humans use (and minerals are used in the production of nearly everything we consume, not just the obvious things such as steel in cars) were laid down primarily during the formation of the Earth. Making new minerals is currently impossible. Alchemists of old tried futilely to make gold, and it still can't be done.

Nearly all the minerals we use are obtained by mining. All mining damages the environment, but probably none more so than strip mining. In strip mining, layers of the Earth are removed in layers in order to gain access to the minerals. It's more economical than tunnel mining, but it damages the land, often irreparably. Strip mining also creates a made-to-order site for erosion and runoff of toxic chemicals.

Fossil fuels present a double whammy for humans and the biosphere. First, they are definitely unrenewable. The coal, oil, and gas deposits we use to fuel most of our power and transportation systems were laid down during the Carboniferous period roughly 300 million years ago. The Earth as we know it is no longer creating fossil fuels as we know them. Second, the use of these fossil fuels as humans currently use them is destroying the biosphere through pollution.

Think About It!

To counteract the toxic effects of fossil fuels, countries around the world are turning to alternative sources of energy. The majority of Sweden's power comes from hydroelectricity, generated from the force of its many rivers and streams. Large portions of New Zealand rely on geothermal energy, generated by enormous hot springs, similar to those at Yellowstone National Park. For more information on hydroelectric power, visit the Web site http://www.soton.ac.uk/~engenvir/environment/alternative/hydropower/hydrenvi.htm.

Pollution and Global Climate Change

Pollution. It's a word we all know and recognize—or do we? What exactly is *pollution*? There are all kinds of pollutants, but scientists group them into two main categories: biodegradable and nonbiodegradable. Biodegradable pollutants can be decayed by microorganisms; nonbiodegradable pollutants can't.

Bio Buzz

Pollution is an undesirable change in the biological, chemical, or physical characteristics of any ecosystem caused by humans.

Think About It!

When you see a brown haze hanging over a large city, such as Los Angeles or Mexico City, it's photochemical smog. The photochemical smog found in Mexico City is so pervasive and toxic that breathing the air there is the equivalent of smoking two packs of cigarettes a day.

Pollutants are also classified as either primary or secondary pollutants. Any pollutant that is emitted directly into the atmosphere is a primary pollutant. Most primary pollutants are dangerous to humans, although the damage may not be readily apparent. Others are immediately damaging. Carbon monoxide, for instance, is present in the exhaust of automobiles. It quickly combines with hemoglobin and affects the ability of the blood to carry oxygen. Lead, hydrocarbons, and sulfur dioxide are some other primary pollutants.

Particulates, the tiny solid particles that are found in smoke and exhaust emissions, are especially damaging to humans. When inhaled, they lodge in the lungs, creating irritation and causing chronic lung diseases.

Secondary pollutants occur as the result of something—frequently something that naturally occurs in the biosphere—acting upon a primary pollutant. Sunlight creates a series of chemical reactions with primary pollutants in the atmosphere, thus producing even more secondary pollutants. When primary pollutants are exposed to sunlight, it creates a secondary mix called photochemical smog.

Besides being toxic to humans, photochemical smog damages the photosynthetic tissues of plants. If plants can't engage in photosynthesis, less oxygen is produced for the rest of us.

Other primary pollutants, such as nitrogen oxide, which is a byproduct of combustion engines, react with sunlight and other chemicals in the atmosphere to form ozone, another type of secondary pollutant.

Acid Rain

Sulfur and nitrogen oxides are among the worst air pollutants. Factories, metal smelters, and coal-burning power plants are the prime emitters of sulfur dioxides. Nitrogen-rich fertilizers, gas- and oil-burning power plants, and automobiles are the main emitters of nitrogen oxides. When these chemicals dissolve in the water that's present in the atmosphere, they form weak solutions of nitric and sulfuric acids.

Winds can carry these acidic solutions enormous distances. When they fall to Earth in the form of rain or snow, we call it acid rain.

Normal rainwater has a pH of approximately 5. The pH of acid rain can be anywhere from 10 to 100 times more acidic. It's not uncommon to have acid rain be as acidic as lemon juice. This acid eats away at marble buildings, rubber, metal, plastics, and other materials—not to mention damaging entire ecosystems.

Some ecosystems can handle acid rain better than others. Highly alkaline soil, or water with a high carbonate content, will neutralize acid. However, many regions in the northern Europe and southern Canada, as well as some scattered regions in the United States, have a thin layer of soil overlaid on solid granite. These ecosystems cannot buffer the acid, and they suffer greatly.

Long ago, researchers confirmed that emissions from motor vehicles, power plants, and factories were the prime sources of air pollution. Further studies indicate that if you live in a city with dirty air, you can easily whack a year off your life span, particularly if the pollution contains particulates.

Weird Science

In eastern North America, the rain has become 30 to 40 times more acidic than it was a few decades ago. More than 200 lakes in New York have already seen their fish populations vanish. Many scientists predict that fish also will disappear from 48,000 lakes in Ontario, Canada, within the next two decades. Crop yields are diminishing in these regions, and forest trees are dying, along with the mycorrhizae that make new growth possible.

Just cleaning up your own region is no guarantee that the pollutants and acid rain will disappear. Most of the air pollutants in Scandinavia, Switzerland, Austria, and the Netherlands comes not from their own country, but from the industrialized regions of Europe. Wind knows no boundaries.

Ozone Thinning

The ozone created by secondary pollution is different from the natural ozone layer that protects the Earth from harmful ultraviolet radiation. The ozone layer that protects us is located almost twice as high above sea level as the peak of Mount Everest. Every September through mid-October, the layer gets thinner at high latitudes, the polar regions. In 1995, the ozone thinning above Antarctica covered an area twice as large as Europe. Thinning at the high northern latitudes already exceeds 10 percent; 60 years from now, scientists believe that the protective ozone layer will be reduced by 30 percent or more over the heavily populated regions of North America, Europe, and Asia.

What's causing this thinning of the ozone layer, and should we even care about it? Chemicals are causing the thinning of the ozone layer, and we should care about it because the ozone layer safeguards our health.

Chlorofluorocarbons (CFCs) are the key chemicals causing ozone thinning. CFCs are compounds of chlorine, fluorine, and carbon. They are odorless and invisible. You can find them in solvents, refrigerators, and air conditioners (as coolants), and in plastic foams. CFCs slowly escape from these products and enter the atmosphere, where they're very resistant to breakdown. Because of the chemical bonding of CFCs, a single atom of chlorine (a component of any CFC) can convert 10,000 or more ozone molecules into oxygen. Before you know it, there's no ozone layer left. And without the ozone layer, we'll most likely see a dramatic rise in skin cancers, weakened immune systems, and eye cataracts.

The ozone layer is thinnest over Antarctica because that's where the ice clouds ultimately end up and melt. Ice provides a surface for the breakdown of chlorine compounds. Ice clouds form in other parts of the globe, but rotating winds around the South Pole pull down the clouds and isolate them from the other latitudes. When the air warms, the chlorine is free to destroy the ozone layer.

Changes in the Earth's atmosphere are also causing temperature variations. The Earth's climate has always been subject to fluctuations in temperature—just look at the fossil record. But currently, the Earth appears to be in a phase of rising temperatures, which is bad news for most organisms around the world. The Earth's temperature has warmed before, naturally, but this current warming phase appears to be almost entirely anthropogenically driven, which means that the warming of the Earth and its seas is being caused by humans.

The current warming of the Earth is referred to as global warming. Not all scientists agree on the causes or severity of global warming, but the majority believe that the burning of fossil fuels is altering the protective layer of atmosphere that surrounds our planet. Sunlight penetrates the atmosphere, warming the Earth and the oceans, but excess heat cannot escape because human-made chemicals have altered the atmosphere. Over time, our Earth is warming and ocean temperatures are rising.

Weird Science

Many organisms are susceptible to temperature changes, but none more so than coral reef colonies. Rising water temperatures stress coral and cause them to eject their zooxanthellae (the symbiotic algae that provide them with food), thus bleaching themselves. Temperatures of 86° to 87.8° are triggers for bleaching. Many corals already live in waters that are very near their upper temperature limit, so a change of only a degree or two can be deadly.

Population Control

If the burning of fossil fuels, global warming, ozone thinning, acid rain, depletion of resources, and habitat destruction weren't enough to keep you up at night, there's always the problem of population control to ponder.

Human population growth may be the single greatest threat our planet faces. Every two weeks, enough people are born to make another Los Angeles; every four weeks, another New York City could be populated. This staggering population growth continues even though at least two billion people are already malnourished or starving, and have no clean drinking water or adequate sanitation facilities. And it continues in regions that are already overcrowded.

Even if by some miracle we could double our current food supply to keep pace with the growth, we'd provide only a marginal standard of living for all. About 20 million to 40 million people would still die from starvation each year. It took 2.5 million years for the first humans to reach the one billion mark. But it only took 45 years to reach the second billion mark, 30 years to reach the third, 15 years to reach the fourth, and 12 years to reach the fifth billion mark. Do you see a trend here? How did humans achieve such explosive growth?

Time	Estimated World Population	Doubling Time (Years)
By 8000 B.C.E.	5 million	1,500
1650 C.E.	500 million	200
1850	1 billion	80
1930	2 billion	45
1960	3 billion	30
1975	4 billion	15
1987	5 billion	12
2050	10 billion (projected)	

Three plausible explanations help to understand the expansion of the human population.

1. We steadily developed the ability to expand into new habitats and climate zones.

2. We increased the productivity of the habitats we were already in.

3. We managed to overcome previously limiting factors.

Humans have done all three of these things. When we learned how to harness fire, build shelters, make clothing and tools, and organize community activities, such as hunts, we were able to move out of the savannah of Africa and into other regions previously unobtainable. About 11,000 years ago, for instance, we moved from a society of hunter-gatherers, to a society of farmers. Instead of following the herds, we stayed in one place, planted and harvested seeds, and domesticated animals.

These two things alone permitted human population growth to increase. New regions opened up for settlement, and better agricultural practices made it easier to grow food. Transportation systems grew up, which also allowed for the distribution of food.

Think About It!

The best way to measure global population trends is through the total fertility rate. This is the average number of children born to women during their childbearing years. In 1950, the average was 6.5 children per woman. In 1996, it was 3. That's an impressive drop, but it's still far above the replacement level of 2.5 per woman. You might think the replacement level would be 2, but you have to take into account that some female children will die before being able to reproduce.

But the human population was still held in check by the limiting factors of disease, poor nutrition, and poor hygiene. Even as recently as 300 years ago, epidemics were still wiping out between one-third and one-half of the population on a fairly consistent basis. Overall, the death rates always exceeded the birth rates.

That's not so any more. With the advent of plumbing and sanitation, humans were able to bring most disease under control. The development of antibiotics, vaccines, and other medicines also eliminated many pathogens. When people discovered how to use fossil fuels, industrialization really took off, as did improved living conditions. The birth rate began to exceed the death rate, and it shows no sign of reversing.

All species face limits to growth, and humans are no exception. Based on what we know of population dynamics and the principles governing population growth, we can expect a dramatic increase in death rates unless new technology breaks us through to a new carrying capacity. Even now, the largest cholera epidemic of the twentieth century is still sweeping through Asia, killing at least five million people. Who knows what the final count for AIDS will be.

Many nations have already begun to implement family-planning measures. China has already established the world's most comprehensive family-planning measures. Other nations are trying to follow suit.

Maintaining the Biosphere

The planet is our home, but it's the biosphere that permits life as we know it to exist. The ocean, the atmosphere, and the land masses interact in ways that we are only just beginning to understand. We do know that these interactions keep us alive.

In the face of pollution, population growth, and the rampant destruction of ecosystems, how can we hope to maintain the biosphere and still preserve our quality of life? Maybe we can't do both. It will take decades, and more likely centuries, to reverse the effects of the damage that we have already inflicted on the biosphere. And not everyone is ready to make these changes. Unfortunately, scattered efforts by individuals are not enough; a united effort by all the individuals of all nations will be needed to reverse trends (and damage) already in motion.

Next to population control, fossil fuels present one of the largest hazards to our planet. If we want to preserve our biosphere, we'll need to find alternative energy sources. Several alternative energy sources are currently being developed. Of these, solar-hydrogen energy, wind energy, and fusion power appear to be the most promising.

Solar-Hydrogen Energy

Every year, the sunlight coming into our planet creates 10 times more energy than all the known fossil fuel reserves. That's approximately 15,000 times as much energy as the world currently uses. Sunlight is powerful, and many people think that we should be using it for something besides crop production.

Bio Buzz

A **photovoltaic cell** is a device that converts solar energy into electricity.

When the electrodes in *photovoltaic cells* are exposed to sunlight, they produce an electrical current that splits water molecules into hydrogen and oxygen gas, which can be used directly as fuel or to generate electricity. Unlike fossil fuels, sunlight and seawater are abundantly available and don't pollute the biosphere. And this isn't new technology—space satellites have been running on this type of fuel for years.

So why aren't we putting resources into developing this energy source? Because fossil fuels are currently cheaper, but that may not be true for long. The largest supplier of natural gas in the United States recently joined forces with a manufacturer of photovoltaic cells. They intend to develop a giant solar facility in the Nevada desert that will be capable of supplying the energy needs of a city of 100,000, for less than the cost any electricity generated by fossil fuels. Hopefully, this plan will succeed and will revolutionize the energy industry.

Wind Energy

Wind has the potential to provide a large portion of our energy needs—and like solar energy, it's nonpolluting. Because winds don't blow on any regular schedule, wind energy isn't as reliable as solar energy (even if the sun isn't shining, the ambient light still produces energy). Even so, California is already getting about 1 percent of its energy from wind farms, and scientists are currently studying the feasibility of wind farms in North and South Dakota. The Dakotas receive so much wind that it's possible that wind farms located there could meet all but 20 percent of our energy needs.

Fusion Power

In recent years, nuclear energy has received quite a lot of negative attention. It certainly can produce a lot of pollution-free power, but it also carries some safety concerns and waste disposal problems that are not associated with other energy production. Due to advancements in this field of science, though, scientists are currently working on creating technology that would mimic the sun's molecular behavior.

The new technology does not have the same safety and waste disposal problems of standard nuclear technology. Fusion power involves confining fuel—in this case, a heated gas composed of two hydrogen isotopes—within magnetic fields and then zapping it with lasers. The fuel implodes and is compressed to incredibly high densities, and energy is released. The stronger the lasers, the greater the compression, and the more fuel is produced. Although the amount of fuel released is increasing, it looks like it will be about 50 years before fusion power will be able to produce enough energy to meet our needs. On the other hand, this is just about when fossil fuels will start running out.

How we manage our population growth and ever-increasing energy needs will be the greatest challenge of the next century. Luck has nothing to do with it when it comes to maintaining our biosphere. Commitment and hard work will preserve out planet and our quality of life.

We've talked about a lot in this chapter, most of it downright depressing. We've discussed how fossil fuels, pollution, ozone thinning, and human population growth are damaging our planet. But we've also talked about some of the new technology that's being developed for alternative energy sources, as well as some simple things individuals can do to promote the health of our planet. And remember: Populations, no matter how large, are still composed of individuals. It's the individuals that make a difference.

The Least You Need to Know

➤ All aspects of our biosphere—the atmosphere, the ocean, and the land—interact in ways that control the living conditions on our planet.

➤ Renewable resources, such as forests, wildlife, and soil, can be managed for long-term benefits. Nonrenewable resources, such as fossil fuels, minerals, and water, cannot be sustained.

➤ Pollution is causing the ozone layer of our atmosphere to thin, acid rain to fall, the Earth and the ocean to warm, and diseases to develop in humans and other organisms.

➤ Human population is expanding at an explosive rate. Every two weeks, enough people are born to create a new Los Angeles.

➤ To maintain our biosphere and life as we know it, we will have to control population growth and develop new sources of energy aside from fossil fuels.

Glossary

acquired immunity   An immunity acquired after birth.

active transport   A method of moving molecules across the plasma membrane. Protein moves against the flow and deliberately moves specific things through the cell membrane.

adaptive radiation   The expansion of a lineage into new habitats and environments through bursts of microevolutionary behavior.

alveoli   A cluster of tiny air sacs at the end of each bronchiole.

ameboid movement   A form of cytoplasmic streaming; the internal flowing motion of the cytoplasm and cell contents.

anatomy   The study of the structure of an organism's body.

angiosperm   A flowering plant.

animalia   Any eukaryotic, multicellular, heterotrophic organism that obtains nutrients through ingestion.

antheridia   The collective male sexual organs of plants.

anthropoid   A primate, characterized by the presence of an opposable thumb and a large cranial capacity relative to body size.

antibiotic   A chemical produced by microorganisms that is capable of inhibiting the growth of many types of bacteria.

antibody   A blood protein that destroys antigens.

antigens   A chemical substance that stimulates the production of antibodies.

apical meristem   A localized region of embryonic, self-perpetuating cells that produces lengthening growth in plants.

archegonia   The collective female sexual organs of plants.

arteries   The part of the human circulatory system that carries oxygenated blood.

asci   Reproductive cells of the Phylum Ascomycota.

ascocarps   The saclike reproductive structure of the Phylum Ascomycota, which contains reproductive cells (asci).

asexual reproduction   A method of reproduction that can take many forms. All forms produce offspring that arise from a single parent and contain only the genes of that parent.

atmosphere   All the air surrounding the Earth. Today, it is mostly nitrogen, hydrogen, and oxygen.

atom   The smallest particle of an element or substance.

autotroph   An organism that uses energy, usually from light, to produce organic molecules from inorganic substances.

auxin   A plant hormone that influences growth by regulating cell elongation.

axon   The long, cylindrically shaped extension of a neuron, or nerve body.

background extinctions   The normal rate of single-species extinctions, as viewed through the fossil record and sedimentary layers across time.

bacteriophage   A virus that infects bacteria. Also referred to as a phage.

bark   All the tissues that are external, or on the outer side, of the vascular cambium in plants.

basidia   The club-shaped sexual spores found in the Phylum Basidiomycota.

basidiocarp   The short-lived reproductive structure found in the Phylum Basidiomycota.

Big Bang theory   The theory that the solar system, and life as we know it, got its start from either one huge explosion or a series of large explosions that occurred in space billions of years ago.

bile   A secretion of the liver that consists of bile salts, bile pigments, cholesterol, and lecithin.

biochemical pathway   A series of complex chemical reactions. Biochemical pathways produce chemical products that are reused. Photosynthesis and respiration are biochemical pathways.

biogenesis   The theory that living organisms come only from other living organisms.

biology   The branch of science that deals with living things. Biology encompasses all aspects—the origin, history, characteristics, habits, and so on—of plants and animals. Biology is also called life science.

bioluminescence   The ability of an organism to produce light.

biome   A geographic area that can be characterized by the specific animals and plants that live there, as well as the climate conditions that exist.

binary fission   The asexual, mitotic division of a unicellular organism that produces identical offspring.

blade   The part of an algae that is leaflike.

blood   A liquid connective tissue comprised of water, solutes, and uniquely formed elements such as platelets and blood cells.

bone   Any part of the hard tissue that forms the skeleton of a vertebrate.

bone tissue   Tissue that consists of collagen fibers and living cells in a ground substance. The cells, fibers, and ground substance are hardened by deposits of calcium salts.

bronchioles   The ending tubes of bronchi.

bronchus   One of two tubes (plural, *bronchi*) of the lower part of the human trachea. Each bronchus leads to a lung.

calorie   A unit of measurement for heat energy.

cardiac cycle   A sequence of contraction and relaxation of the heart muscles.

catastrophism   The doctrine that creation occurred several times, only to be wiped out by a flood each time.

cell cleavage   A part of mitosis in which the cell pinches and divides.

cell cycle   The many repetitions of a cell's growth and reproduction.

cell wall   The rigid covering of a plant cell.

cellular respiration   The breakdown of glucose within the cell.

cerebral cortex   A thin surface layer of cells covering your cerebral hemispheres.

chordate   An animal that, at some stage of its development, has a notochord, a nerve chord, and a muscular pharynx.

chromosome   The curled and condensed form of chromatin. Chromosomes are formed almost entirely of DNA and contain all the genetic information of the cell.

coacervate   An irregularly shaped droplet, formed by a group of molecules of different types.

codominance   An instance in which two or more alleles influence the phenotype of an organism. Also called incomplete dominance.

codon   A specific group of three sequential bases of messenger RNA (mRNA). Codons recognize and attract specific amino acids using transfer RNA (tRNA) as an intermediary.

coelenterates   A name for cnidarians because they possess a coelenteron, or "hollow gut" body pattern.

coenocytic fungi   A fungi that lacks septa within its hypha.

coevolution   When two species jointly change, in close interaction with each other.

cohesion-tension theory   A theory that states that the water in plant xylem is pulled upwards to the leaves by the drying power of evaporation.

colloid   A mixture of particles that don't settle over time.

companion cells   Specialized parenchyma cells that help load organic compounds into the sieve tubes of plants.

comparative morphology   The study of an organism's development, compared to the development of other organisms.

compartmentalization   The process in which a tree deploys compounds and walls off an invader.

complete digestive system   A system that takes in food through an opening at one end and disposes of wastes through an opening at the other end. In between, the food travels through a tube that is divided into specialized regions for food transport, processing, and storage.

complex tissues   Plant tissues that are composed of a variety of cell types.

compounds   Atoms of two or more elements that are joined by a chemical bond. Compounds are either organic (containing carbon and derived from living things) or inorganic (derived from nonliving things).

conidium   An asexually reproduced spore found on the hypha of ascomycetes.

contractile protein   A protein (one or more strands of amino acid) that is capable of responding to stimuli, or contracting.

convergent evolution   An instance in which two unrelated species become more similar in appearance as they adapt to the same environment.

cotyledon   A seed leaf.

crossing-over   An exchange of alleles between homologous chromosomes.

culture   The sum total of all the behavioral patterns of any given social group, handed down from one generation to the next through learning and symbolic behaviors, including writing and language.

cuticle   A waxy, waterproof layer of epidermal cells that covers plant regions that are exposed to air.

cytoplasm   The inner region of a cell.

daughter cells   The end product of mitosis.

decomposer   Any organism that gets its nutrients from dead plants and animals.

development   The process in which specialized, morphologically different body parts emerge.

diastole   The relaxation phase of the cardiac cycle.

diatom   Any species in the large group of microscopic algae that float in the ocean.

dicot   A flowering plant with two cotyledons in its seed, net-veined leaves, and flowers arranged in fours, fives, or multiples thereof.

diffusion   The random movement of molecules across the cell membrane from areas of high concentration to low concentration.

divergent evolution   The process by which two or more related species become increasingly dissimilar.

DNA   The abbreviation for deoxyribonucleic acid. DNA consists of two long strands of nucleotides that are twisted together in a double helix pattern. DNA stores and uses information to direct the cell's activities and replicates itself to create new cells.

DNA library   A collection of DNA fragments, all produced by plasmids and restriction enzymes.

dormancy   A state of decreased metabolic activity.

double fertilization   A process that is unique to angiosperms, involving two sperm. One sperm fertilizes the ovule and forms a zygote. The other sperm fuses with nuclei in the ovule, creating a food supply for the developing embryo, called the endosperm.

double helix   The unique pattern of DNA, similar to a spiral staircase.

early wood   Wood that is composed of the first xylem cells produced during the growing season.

ecosystem   An ecological unit. Each ecosystem is an association of organisms and their physical environment, all of which is interconnected through a continuous flow of energy and material cycling.

element   Any substance that cannot be separated into different substances, except by nuclear reaction or radioactive decay.

embryo   An organism in the earliest stages of development.

endangered species   Any species that is currently at risk of extinction in its native habitat.

endocrine glands   Glands that release their secretions directly into the bloodstream.

endosymbiosis   The theory that prokaryotic intracellular parasites evolved into various cell organelles, forming eukaryotic cells in the process.

epithelial   Cells that adhere closely to one another and are organized in layers.

eukaryotic cell   A cell with a plasma membrane, cytoplasm, and a nucleus. Within the cytoplasm are organelles.

eutrophic lake   A lake that is rich in organic matter.

evolution   The process by which species change over time, evolving into different forms.

exocrine glands   Glands that release their secretions through ducts.

external respiration   The exchange of gases between the atmosphere and the blood.

Fick's Law   A statement that at any given time, the diffusion of gases across the respiratory membrane is dependent upon the surface area of the membrane and the partial pressure across it.

fossil   The hardened remains of a plant or animal from a previous geological age that has been preserved in the Earth's crust.

fossil record   A term scientists use to refer to the layering of the Earth's crust and the fossils that have been preserved within the layers.

fungal spore   A walled cell or multicellular structure that is produced either sexually (through meiosis) or asexually (through mitosis).

fungi   Heterotrophs that survive by decomposing living or nonliving organic matter.

gametangia   The gamete-producing structures that develop within hypha.

gamete   A sexual reproductive cell.

gametophyte   A multicellular structure found in plants and algae that produces gametes, or sexual reproductive cells.

genome   The complete set of instructions for making an organism. The human genome is all DNA found in a human being, including its genes.

genotype   The genetic makeup of an organism.

germ layers   The layers of cells that originate in the developing embryo and become specific structures in the animal.

gills   Organs specializing in the exchange of gases with water.

glycolysis   The breakdown of glucose into pyruvic acid.

growth   The process by which the number, size, and volume of the cells of multicellular organisms increase.

gymnosperm   A vascular plant that bears seeds on the exposed surfaces of reproductive structures, such as cone scales.

half-life   The period of time it takes for half of an unstable isotope to decay.

helper T-cell   A type of T-cell that can produce interleukins and stimulate B-cells to reproduce clone cells.

hemoglobin   An iron-containing protein found in blood.

herbaceous plant   Any seed plant whose stem withers and dies on a yearly basis.

heredity   The transmission of traits from parents to their offspring.

hermaphrodite   An organism that produces both eggs and sperm.

heterospory   The process in which two types of spores are produced. Conifers and seed plants both produce two types of spores.

heterotroph   An organism that cannot produce its own energy.

holdfast   A cellular part of algae that anchors it to rocks and other objects.

homeostasis   The process by which cells regulate and control their own environment.

homologous structures   Structures that resemble one another in body form or pattern across all species.

hormones   Substances that are produced in one part of the body and that influence the activity of cells in another part of the body.

hypha   A filament of the mycelium.

hypoxia   Cellular oxygen deficiency, a condition that affects visitors to high altitudes.

immune   To be exempt from or protected from something disagreeable or harmful.

incomplete digestive system   A digestive system in which the gut has only one opening.

inheritance   The passing of traits through heredity.

integument   To cover; the system we call skin.

integumentary exchange   The process in which the respiratory system, oxygen, and carbon dioxide simply diffuse across the thin, moist layer of surface epidermal cells.

internal respiration   The exchange of gases between the blood and the cells of the body.

internode   The stem region between two nodes.

interphase   The first step of cell division.

interstitial fluid   The extracellular fluid that occupies the spaces between cells and tissues.

invertebrate   An animal without a backbone or spine.

labeled DNA probe   A short DNA sequence that has been synthesized from radioactively labeled nucleotides.

late wood   Wood that is composed of xylem cells formed further into the growing season.

lateral meristem   A localized region of embryonic, self-perpetuating cells with growth that increases the circumference of a plant.

lenticils   Localized areas in the cork layer where the cells are loosened up a bit. Gases are able to pass through these open areas of the cork layer.

lichen   A symbiotic relationship between a fungus and a green algae, or another photosynthetic partner.

light-dependent reactions   A two stage process. The first stage converts solar energy into electrical energy. The second stage converts the electrical energy into chemical energy.

lineage   An organism's line of descent, or ancestry.

linkage group   A group of genes located on the same chromosome. Genes in a linkage group usually are inherited together.

loams   Soils with roughly equal proportions of sand, silt, and clay.

lung   An internal respiratory surface, shaped like a sac or a cavity.

lymph nodes   Small organs that filter lymph, primarily located in your groin, neck, and armpit areas.

macroevolution   The process of evolution on a grand scale. Macroevolution looks for rates of change, patterns, and trends among whole groups of species.

macrophage   A phagocytic white blood cell that develops from circulating white blood cells and that defends body tissues.

mammal   An animal that gives birth to live young, nurses its young, maintains a constant body temperature, and has hair on its body.

mantle   The organ that secretes a shell in mollusks.

mantle cavity   In mollusks, the space between the mantle and the visceral mass. Gills usually are located in the mantle cavity.

mass extinctions   The process of whole major groups of species becoming extinct at the same time.

matter   What a thing is made of.

meiosis   The nuclear division process that reduces the number of chromosomes in any given cell by half.

membrane excitability   When a cell produces a nerve impulse in response to stimulation.

menstrual cycle   A cycle that usually occurs in females in cycles of 28 days. During that time, an ovum matures and is positioned to meet a sperm in the Fallopian tube. The ovum travels through the Fallopian tube into the uterus. If fertilization does not occur, the egg will be discharged through the vagina.

meristem   A localized region of embryonic, self-perpetuating cells found in plants.

mesophyll   The ground tissue of plant leaves, composed of chloroplast-rich parenchyma cells.

metabolize   To utilize energy.

metamorphosis   The changes by which an organism passes through into adulthood.

microevolution   The study of evolution on a small scale, as it relates to a single species.

microfossil   A fossilized cell or chain of cells that can be seen only with a microscope.

microsphere   A spherically shaped droplet that is typically formed by a group of molecules of the same type.

molecule   The smallest part of an element that can exist by itself and still retain the characteristics of the element or substance.

monera   Prokaryotic organisms that lack a nucleus and other membrane-bound organelles.

monocot   A vascular flowering plant with only a single cotyledon, or seed-leaf, in its seed, parallel veined leaves, and flowers that usually come in multiples of three.

monohybrid cross   A cross between individuals that involves one pair of contrasting traits.

morphology   The internal and external appearance of an organism.

mucosa   The innermost layer of your gut.

muscular pharynx   The entrance to an organism's digestive tract; frequently an entrance to the respiratory tract as well.

mutation   The process by which a molecule of DNA alters from its original pattern.

mycelium   The matted, branching, filamentous network of a fungi.

mycorrhiza   A symbiotic relationship in which fungal hyphae associate with plant roots.

myofibril   The contracting, threadlike structure of a skeletal muscle.

myth   A story that presents itself as a factual account of a true experience. Myths take place beyond the time of anything that ever was.

natural immunity   An immunity that is present at birth.

natural selection   The process by which organisms with favorable variations survive in the natural environment and reproduce more individuals at higher rates than those organisms without the advantageous variations.

neritic zone   The subarea of the ocean that extends from a shoreline over the continental shelf.

nerve chord   A hollow tube located just above the notochord; a chordlike communication line consisting of nerve cells.

nerve impulse An abrupt but brief reversal in the steady voltage difference that exists in a resting neuron and some other cells.

nerve net A type of nervous system found in cnidarians.

neuroglial cells Cells found in vertebrates that provide structural and metabolic support for neurons.

neuron The basic unit of communication within your nervous system.

nitrogen-fixing The process of converting gaseous nitrogen (N_2) into ammonia compounds.

node The place on a plant stem where the leaf attaches.

nonspecific defense mechanism A physical, chemical, or biological barrier that serves as a shield against a wide range of pathogens.

nonvascular plants Plants without true vascular tissues, true leaves, roots, or stems.

notochord A dorsal chord of specialized cells, similar to a spine.

nuclear transfer The process by which the nucleus is removed from an egg cell and is replaced with the nucleus from a cell of the organism to be cloned.

nucleoid In prokaryotic cells, the portion that contains the genetic material.

nucleotides A small molecule of RNA and DNA, consisting of a nitrogen base, a sugar, and a phosphate group.

nucleus Located within the cell's cytoplasm, the nucleus encloses the cell's genetic material.

nutrient Any element that is essential to an organism because, either directly or indirectly, that element plays a role in the organism's growth and survival that no other element can fulfil.

nutrition The process by which an organism takes in and assimilates food, or anything that nourishes.

obligate intracellular parasite A parasite that can reproduce only by invading a host cell and using that cell's enzymes and organelles to produce more viruses.

oligotrophic lake A lake that contains very little organic matter.

organelles Membrane-bound regions of a cell that enclose specific chemical activities.

organic compounds A compound made up of organic elements, those elements that are needed to form life, such as carbon, hydrogen, or oxygen.

osmosis The random movement of water molecules across a cell membrane from an area of high concentration to one of low concentration.

ovary The primary female reproductive organ that is responsible for the formation of eggs.

ovule   The single plant cell from which all growth starts.

ovum   The single animal cell from which all growth starts.

oxidation   The process in which a molecule unites with oxygen, as in burning or rusting.

oxidizing atmosphere   An atmosphere that is rich in oxygen.

paleontology   The study of prehistoric life through fossils.

parasite   An organism that survives by getting its nutrients from a living host.

pathogen   Any organism that causes disease.

periderm   The major tissue of bark.

phagocytosis   The ingestion of nutrients by engulfing.

phloem   The tissue that carries sugars made through photosynthesis from the leaves to the rest of the plant.

photosynthesis   A series of chemical reactions in plants that convert radiant energy from the sun into chemical energy that the plant can use.

photovoltaic cell   A device that converts solar energy into electricity.

physiology   The study of living organisms and the ways they carry out the functions necessary for life.

placenta   A blood-engorged organ that develops in pregnant female mammals. The placenta permits exchanges between the mother and developing fetus without intermingling their blood.

Plantae   The kingdom that consists of all multicellular and autotrophic plants.

plasma membrane   The cell's outermost layer.

plastid   An organelle that converts solar energy into chemical energy.

pollution   An undesirable change in the biological, chemical, or physical characteristics of any ecosystem.

pressure flow theory   A statement that internal pressure in the plant builds up at the source of the sieve tube system and then pushes the nutrient-rich solution on toward a sink, where the nutrients are unloaded.

primary immune response   The team effort of your B-cells producing antibodies and your T-cells attacking directly.

primate   A specific class of mammals. Most primates have highly movable fingers and toes that have flattened nails instead of claws, good vision, a larger cranial capacity, and the ability to hold their bodies upright.

Principle of Dominance and Recessiveness   Developed by Gregor Mendel, a theory stating that one factor in a pair may mask the other factor, preventing it from having any effect.

Principle of Independent Assortment Developed by Gregor Mendel, a theory stating that factors for different traits are distributed independently in reproductive cells.

Principle of Segregation Developed by Gregor Mendel, a theory stating that the two factors for a trait segregate during the formation of sperm and eggs.

prokaryotic Meaning "before the nucleus." Often refers to bacteria that existed on Earth before the evolution of cells with a nucleus.

prophage A short piece of DNA that enters a host's cells and then just sits there doing nothing.

protein synthesis The formation of proteins using information that is coded upon DNA and carried out by RNA.

proteins Complex organic compounds formed from amino acids. Protein forms the basis and structure of a cell.

protists Eukaryotic organisms with nuclei and membrane-bound organelles, but lacking organized tissue systems.

pseudopodia Cytoplasmic extensions that enable protozoans to move.

pulmonary circuit The movement of blood from your heart into your lungs, and then back to the heart.

radioactive Something that has an unstable molecular structure and that gives off particles as it disintegrates.

radioactive isotope An element with an unstable molecular structure, which causes it to decay and form into other elements.

recombinant DNA technology Technology that involves procedures in which genes in DNA from different species are isolated, cut, and spliced together, and in which the new recombinant molecules are then reproduced in quantity.

reducing atmosphere An atmosphere with little or no oxygen present.

renewable resource Any material that can be regrown or replenished.

respiration The release of chemical energy for cellular functions.

resting membrane potential The steady voltage difference across the plasma membrane of a resting neuron.

RFLP An acronym for "restriction fragment length polymorphisms," otherwise known as DNA fingerprinting.

rhizoid A rootlike structure that attaches a gametophyte to rocks, soil, or tree bark.

rhizome The underground stem of a fern.

ribonucleic acid A single strand of nucleotides, responsible for the production of proteins in a cell.

Geological Time Scale of Life

End Date (millions of years ago)	Era	Period	Epoch	Organisms
0.1	Cenozoic	Quaternary	Recent	Modern humans appear
			Pleistocene	Woolly mammoths appear
2.5		Tertiary	Pliocene	Apes and large carnivores appear
7			Miocene	Land mammals diversify
26			Oligocene	Horses and primitive apes appear
38			Eocene	Small horses appear
53			Paleocene	Primates and the first carnivores appear
65	Mesozoic	Cretaceous		Dinosaurs die out; flowering plants appear
135		Jurassic		Age of the dinosaurs; first birds arise
195		Triassic		First dinosaurs; mammals; and forests of conifers appear
225	Paleozoic	Permian		First seed plants appear
280		Carboniferous		First reptiles appear
345		Devonian		First insects and amphibians appear
395		Silurian		Most life in the ocean, fishes dominant; first modern vascular plants invade land
430		Ordovician		Modern groups of algae and fungi appear
500		Cambrian		First fish; many invertebrates and marine plants appear
600	Precambrian			First eukaryotes; blue-green bacteria and bacteria abound

Index